T0215495

# SpringerBriefs in Mathematics

**SpringerBriefs in Mathematics** showcases expositions in all areas of mathematics and applied mathematics. Manuscripts presenting new results or a single new result in a classical field, new field, or an emerging topic, applications, or bridges between new results and already published works, are encouraged. The series is intended for mathematicians and applied mathematicians.

More information about this series at http://www.springer.com/series/10030

SpringerBriefs present concise summaries of cutting-edge research and practical applications across a wide spectrum of fields. Featuring compact volumes of 50 to 125 pages, the series covers a range of content from professional to academic. Briefs are characterized by fast, global electronic dissemination, standard publishing contracts, standardized manuscript preparation and formatting guidelines, and expedited production schedules.

**Typical topics might include:**

- A timely report of state-of-the art techniques
- A bridge between new research results, as published in journal articles, and a contextual literature review
- A snapshot of a hot or emerging topic
- An in-depth case study
- A presentation of core concepts that students must understand in order to make independent contributions

*Titles from this series are indexed by Web of Science, Mathematical Reviews, and zbMATH.*

Yong Cheng

# Incompleteness for Higher-Order Arithmetic

## An Example Based on Harrington's Principle

 Springer

Yong Cheng
School of Philosophy
Wuhan University
Wuhan, Hubei, China

ISSN 2191-8198          ISSN 2191-8201   (electronic)
SpringerBriefs in Mathematics
ISBN 978-981-13-9948-0          ISBN 978-981-13-9949-7   (eBook)
https://doi.org/10.1007/978-981-13-9949-7

This Springer imprint is published by the registered company Springer Nature Singapore Pte Ltd.
The registered company address is: 152 Beach Road, #21-01/04 Gateway East, Singapore 189721, Singapore

*This book is the research result of Humanities and Social Sciences of Ministry of Education Planning Fund Project and Wuhan University's independent scientific research project (Humanities and Social Sciences). This book is supported by "Ministry of Education Humanities and Social Sciences Research Planning Fund" (project no: 17YJA72040001) and "the Fundamental Research Funds for the Central Universities".*

# Preface

Hilbert proposed his famous list of 23 open problems at the International Congress of Mathematicians in Paris in 1900 [1]. The second problem deals with the question whether the axioms of mathematics are *consistent*, i.e., that no contradiction can be derived. Hilbert later elaborated on the second problem by proposing *Hilbert's program for the foundations of mathematics* [2]. The aim was to find a consistent set of axioms whose consequences comprise all theorems in mathematics.

As is well known, Gödel's first incompleteness theorem shows that for any logical system that can accommodate arithmetic, there are true sentences that cannot be proved. Hence, it is not possible to formalize all of mathematics within a consistent formal system, as any attempt at such a formalism will omit some true mathematical statements. In this light, Gödel's incompleteness theorem is one of the most important theorems in foundations of mathematics and mathematical logic in the twentieth century and has had a huge impact on the development of logic, philosophy, mathematics, computer science, and other fields. In the literature, there are a number of good research books on Gödelian incompleteness (e.g., [3–8]) and a huge number of research articles.

Now, Gödel's true-but-unprovable sentence from the first incompleteness theorem is purely logical in nature, i.e., not mathematically natural or interesting. In this light, an interesting problem is to find mathematically natural and interesting statements that are similarly unprovable. A lot of research has since been done in this direction, most notably by Harvey Friedman. A lot of examples of *concrete incompleteness* with real mathematical content have been found to date. Section 1.1.3 provides an overview of research on incompleteness in higher-order arithmetic.

This book contributes to Harvey Friedman's research program on concrete incompleteness for *higher-order arithmetic*. In a nutshell, I shall introduce the set-theoretic hierarchy $Z_n$ of higher-order arithmetic. Here, $Z_2$, $Z_3$ and $Z_4$ are the corresponding set-theoretical axiomatic systems for second-order arithmetic, third-order arithmetic, and fourth-order arithmetic. I then formulate a concrete mathematical theorem expressible in the language of second-order arithmetic which is neither provable in $Z_2$ or $Z_3$, but provable in $Z_4$. While I do not provide a comprehensive study of incompleteness, Sect. 1.1.3 includes some examples of concrete mathematical theorems about arithmetic which are not provable in PA; examples of concrete mathematical theorems about arithmetic which are not provable in certain sub-systems of second-order arithmetic stronger than PA; and examples of concrete mathematical theorems about analysis provable in third-order arithmetic but not provable in second-order arithmetic.

In this book, I examine the aforementioned Hilbert's program "relativized" to $Z_2$, which deals with the following question: are all theorems in classic mathematics expressible in second-order arithmetic provable in $Z_2$? Now, most classic mathematical theorems about real numbers expressible in (the language of) second-order arithmetic are also provable in $Z_2$. Nonetheless, I shall provide a negative answer to this question in the form of a counterexample which stems from a famous theorem in set theory, namely *the Martin-Harrington Theorem*. The latter expresses between $Det(\Sigma_1^1)$ and the existence of $0^\sharp$, establishing the equivalence between large cardinal and determinacy hypotheses. The Martin-Harrington Theorem is expressible in second-order arithmetic and provable in ZF. In this book, I give a systematic analysis of known proofs of the Martin-Harrington Theorem in higher-order arithmetic.

It is known that Martin's Theorem, i.e., that the existence of $0^\sharp$ implies $Det(\Sigma_1^1)$, is provable in $Z_2$ (cf. Sect. 3.1). However, in the proof of Harrington's Theorem, i.e., $Det(\Sigma_1^1)$ implies the existence of $0^\sharp$, Harrington makes use of a principle nowadays called *Harrington's Principle* (HP hereafter). All known proofs of Harrington's Theorem are done in the following two steps: first prove that $Det(\Sigma_1^1)$ implies HP and then show that HP implies that $0^\sharp$ exists. Below, I show that the first implication "$Det(\Sigma_1^1)$ implies HP" is provable in $Z_2$. A natural question is then whether the second implication is provable in $Z_2$. By way of a negative answer, I show that the statement that the second implication, i.e., HP implies that $0^\sharp$ exists, is not provable in $Z_2$. This provides the required counterexample for Hilbert's program relativized to $Z_2$.

Moreover, I show that "HP implies that $0^\sharp$ exists" is also not provable in $Z_3$, but is provable in $Z_4$. As part of joint work with Ralf Schindler, I prove in Sect. 2.2 that $Z_2 + $ HP is equi-consistent with ZFC. In Sect. 2.3, $Z_3 + $ HP is shown to be equi-consistent with ZFC $+$ there exists a remarkable cardinal. In Sect. 2.4, I show that HP is equivalent to $0^\sharp$ exists in $Z_4$. Hence, $Z_4$ is the minimal system from higher-order arithmetic to show that HP implies that $0^\sharp$ exists.

It is unknown whether the Harrington Theorem is provable in $Z_2$. I show in Chap. 3 that the **boldface** Martin-Harrington Theorem is provable in $Z_2$. In Chap. 4, we examine the large cardinal strength of the strengthening of HP, called HP($\varphi$), over $Z_2$ and $Z_3$. In Sect. 2.3, we force a model of "$Z_3 + $ Harrington's Principle" via class forcing using the reshaping technique assuming the existence of a remarkable cardinal. In Chap. 5, I force a model of "$Z_3 + $ Harrington's Principle" via set forcing without the use of the reshaping technique and assuming there exists a remarkable cardinal with a weakly inaccessible cardinal above it. For the proof of the main Theorem 5.1 in Chap. 5, I introduce the notion of "strong reflecting property" for $L$-cardinals in Sect. 5.2. In Chap. 6, I develop the full theory of the strong reflecting property for $L$-cardinals and characterize the strong reflecting property of $\omega_n$ for $n \in \omega$.

This book is based on my dissertation [9] and sequent work in [10–12]. In general, Chaps. 2, 4, 5, and 6 are revisions and improvements of dissertation and sequent work in [10–12] to fit into the current theme of concrete incompleteness for higher-order arithmetic.

I felt it was necessary for me to write this book for the following reasons. Firstly, this book contributes to the research program on concrete incompleteness for higher-order arithmetic and gives a systematic analysis of the Martin-Harrington Theorem in higher-order arithmetic. In particular, this book gives a specific example of concrete mathematical theorems which is expressible in second-order arithmetic, but the minimal system in higher-order arithmetic to prove it is $Z_4$.

Secondly, this book is a significant expansion and improvement over my dissertation. I have strengthened the main results and filled in some technical gaps in my dissertation. The large cardinal strength of "$Z_2 + \mathsf{HP}$" and "$Z_3 + \mathsf{HP}$" is not discussed in [9], while I establish the exact large cardinal strength of "$Z_2 + \mathsf{HP}$" and "$Z_3 + \mathsf{HP}$" below. Also, the large cardinal hypothesis used in my proof of forcing a model of "$Z_3 + \mathsf{HP}$" via set forcing in Chap. 5 are much weaker than the hypothesis used in [9]. Furthermore, this book contains some new materials not covered in [9–12]. For these reasons, I felt it was necessary to write this book which contains all my current results on the analysis of the Martin-Harrington Theorem in higher-order arithmetic. I will assume that readers are already familiar with the basics of forcing, large cardinals, effective set theory, determinacy, admissible ordinals, and reverse mathematics. However, I will try my best to make this book self-contained.

This book makes a contribution to the foundations of mathematics and may be relevant for philosophers of mathematics and for three of the four major branches of mathematical logic, namely set theory (large cardinals, descriptive set theory, determinacy), recursion theory (admissible ordinals, higher recursion theory, the analytic hierarchy) and proof theory (reverse mathematics), and therefore certainly relevant for mathematical logicians.

Wuhan, China
October 2018

Yong Cheng

# References

1. Hilbert, D.: Mathematical problems. Bull. Amer. Math. Soc. **8**, 437–479 (1902)
2. Hilbert, D.: Über das Unendliche. Mathematische Annalen, **95**, 161–190 (1926)
3. Murawski, R.: Recursive Functions and Metamathematics: Problems of Completeness and Decidability, Gödel's Theorems. Springer, Netherlands (1999)
4. Lindström, P.: Aspects of Incompleteness. Lecture Notes in Logic v. 10 (1997)

5. Smith, P.: An Introduction to Gödel's Theorems. Cambridge University Press (2007)
6. Smullyan, M.R.: Gödel's Incompleteness Theorems. Oxford Logic Guides 19. Oxford University Press (1992)
7. Smullyan, M.R.: Diagonalization and Self-Reference. Oxford Logic Guides 27. Clarendon Press (1994)
8. Hájek, P., Pudlák, P.: Metamathematics of First-Order Arithmetic. Springer, Berlin, Heidelberg, New York (1993)
9. Cheng, Y.: Analysis of Martin-Harrington theorem in higher-order arithmetic. Ph.D. thesis, National University of Singapore (2012)
10. Cheng, Y.: Forcing a set model of $Z_3 +$ Harrington's principle. Math Logic Quart. **61**(4–5), 274–287 (2015)
11. Cheng, Y.: The strong reflecting property and Harrington's principle. Math Logic Quart. **61**(4–5), 329–340 (2015)
12. Cheng, Y., Schindler, R.: Harrington's principle in higher-order arithmetic. J. Symb. Log. **80**(02), 477–489 (2015)

# Acknowledgements

First of all, I thank my fortress, my stronghold, my rock, my shield, my refuge, my deliverer and my strength. Without him, this book would not be possible. I thank my Ph.D. adviser Prof Chong Chi Tat and W. Hugh Woodin. Especially, I am indebted to Prof W. Hugh Woodin for his support and supervision during my Ph.D study; without his supervision the work in my dissertation would not have come out. I thank Prof Ralf Schindler for his support for my two-year postdoctoral work at University of Muenster and his contribution in our joint paper in [1]. I thank all the reviewers for their works and helpful comments. Especially, I thank one anonymous reviewer with a long list of comments for improvement. I thank Leo A. Harrington, Ulrich Kohlenbach, Michael Rathjen, Sam Sanders, Ralf Schindler, Andreas Weiermann and W. Hugh Woodin for their helpful comments on this book. I thank Sam Sanders for his help to proofread and polish the language of this book; without his work, this book would not be in such a good shape. I thank Series Editors of "SpringerBriefs in Mathematics" for their recommendations of this book. I thank Springer editors Ramon Peng, Daniel Wang and AP Umamagesh for their hard work. I thank the publisher Cambridge University Press for the permission to reuse some materials in [1]. I thank the publisher Wiley-VCH for permission to reuse some materials in [2, 3]. Finally, I thank the Chinese Ministry of Education for their support via the Humanities and Social Sciences Planning Fund (project no. 17YJA72040001) and Wuhan University in China for the support from "the Fundamental Research Funds for the Central Universities".

# Reference

1. Cheng, Y., Schindler, R.: Harrington's principle in higher-order arithmetic. J. Symb. Log. **80**(02), 477–489 (2015)
2. Cheng, Y.: Forcing a set model of $Z_3$ + Harrington's Principle. Math. Logic. Quart. **61**(4–5), 274–287 (2015)
3. Cheng, Y.: The strong reflecting property and Harrington's Principle. Math. Logic. Quart. **61**(4–5), 329–340 (2015)

# Contents

# Chapter 1
# Introduction and Preliminaries

**Abstract** In this chapter, I provide an overview of Incompleteness, Reverse Mathematics, and Incompleteness for higher-order arithmetic, respectively in Sects. 1.1.1, 1.1.2 and 1.1.3. This should provide the reader with a good picture of the background and put the main results in this book into perspective. In Sect. 1.1.4, I review some of the notions and facts from Set Theory used in this book. In Sect. 1.2, I introduce the main research problems and outline the structure of this book.

## 1.1 Preliminaries

### 1.1.1 Basics of Incompleteness

In this section, I give a brief introduction to Gödel's incompleteness theorems. In particular, I present different versions of *Gödel's first incompleteness theorem* and *Gödel's second incompleteness theorem*: from the original version to the modern generalized versions.

Gödel's incompleteness theorem is one of the most remarkable and profound discoveries of the 20th century, an important milestone in the history of modern logic, which has had wide and profound influence on the development of logic, philosophy, mathematics, computer science and other fields, substantially shaping mathematical logic and foundations of mathematics after its publication in 1931.

The impact of Gödel's incompleteness theorem is not confined to the community of logic or mathematics. Indeed, Feferman writes the following about the impact of Gödel's incompleteness theorems.

> their relevance to mathematical logic (and its offspring in the theory of computation) is paramount; further, their philosophical relevance is significant, but in just what way is far from settled; and finally, their mathematical relevance outside of logic is very much unsubstantiated but is the object of ongoing, tantalizing efforts. ([1], p. 434).

© The Author(s), under exclusive license to Springer Nature Singapore Pte Ltd. 2019
Y. Cheng, *Incompleteness for Higher-Order Arithmetic*, SpringerBriefs
in Mathematics, https://doi.org/10.1007/978-981-13-9949-7_1

1

In the following, I first review some notions and facts used in this chapter. Notations will all be standard. For books on Gödel's incompleteness theorem, I refer to [2–8]. For survey papers on Gödel's incompleteness theorem, I refer to [9–12]. For meta-mathematics of sub-systems of Peano Arithmetic, I refer to [13].

First of all, a *theory* is a deductively closed set of sentences in a first-order language. For a given theory $T$, we use '$L(T)$' to denote the language of $T$ and often equate $L(T)$ with the list of non-logical constants of the language. For a formula $\phi$ in $L(T)$, let '$T \vdash \phi$' denote that $\phi$ is provable in the theory $T$.[1] We say that '$T$ is consistent' if no contradiction is provable in $T$. We say that '$\phi$ is independent of $T$' if $T \nvdash \phi$ and $T \nvdash \neg\phi$. We say that '$T$ is complete' if for any sentence $\phi$ in $L(T)$, either $T \vdash \phi$ or $T \vdash \neg\phi$; otherwise, $T$ is called 'incomplete' (i.e. there is a sentence $\phi$ in $L(T)$ such that $\phi$ is independent of $T$).

In this book, I always assume that first-order theories $T$ can be *arithmetized* via a recursive set of non-logical constants. By 'arithmetization', I mean that any formula or finite sequence of formulas can be coded by a natural number, and this code is called the *Gödel's number*. Under this arithmetization, we can in principle establish the one-to-one correspondence between formulas of $L(T)$ and natural numbers. Under this correspondence, we can translate meta-mathematical statements about the theory $T$ into statements about natural numbers. Any statement about natural numbers pertaining to recursive relations, is therefore representable in the theory $T$. Consequently, one can speak about the meta-mathematical properties of the theory $T$ *inside the theory $T$ itself*. This is the essence of Gödel's idea of arithmetization. For details of arithmetization, I refer to [3].

Next, a theory $T$ is *recursively axiomatizable* if it has a recursive set of axioms, i.e. the set of Gödel numbers of axioms of $T$ is recursive. A theory is *finitely axiomatizable* if the set of axioms is finite. A theory $T$ is *recursively enumerable* (r.e.) if it has a recursively enumerable set of axioms, i.e. the set of Gödel numbers of axioms of $T$ is recursively enumerable. A theory $T$ is *essentially incomplete* if any recursively axiomatizable consistent extension of $T$ in the same language is incomplete. A theory $T$ is *minimal essentially incomplete* if $T$ is essentially incomplete and if deleting any axiom of $T$, the remaining theory is no longer essentially incomplete.

Let $L_2$ denote the language of second-order arithmetic (**SOA**) which has two distinct sorts of variables: variables of the first sort are number variables, ranging over the set of natural numbers $\omega$, to be denoted by $i, j, k, \ldots$; variables of the second sort are set variables, ranging over all subsets of $\omega$ to be denoted by $X, Y, Z, \ldots$.

Correspondingly, there are two sorts of quantifiers in $L_2$: the number quantifiers and set quantifiers, to be denoted by $\exists n, \forall n$ and $\exists X, \forall Y$ respectively. The terms and formulas of $L_2$ are recursively defined as follows. Terms of $L_2$ are number variables, the constant symbols 0 and 1, and $t_1 + t_2$ and $t_1 \cdot t_2$ whenever $t_1$ and $t_2$ are terms of $L_2$ (terms are intended to denote natural numbers). Atomic formulas of $L_2$ are $t_1 = t_2, t_1 < t_2$, and $t_1 \in X$ where $t_1$ and $t_2$ are terms of $L_2$ and $X$ is a set

---

[1]i.e. there is a finite sequence of formulas $\langle \phi_0, \ldots, \phi_n \rangle$ such that $\phi_n = \phi$, and for any $0 \leq i \leq n$, either $\phi_i$ is an axiom of $T$ or $\phi_i$ follows from some formulas before $\phi_i$ in the list by using one of the inference rules.

variable. Formulas of $L_2$ are built up from atomic formulas by means of propositional connectives, number quantifiers, and set quantifiers.

A model for $L_2$ is an ordered 7-tuple $\mathfrak{M} = (M, \mathscr{S}_M, +_M, \cdot_M, 0_M, 1_M, <_M)$, where $M$ is a set which serves as the range of the number variables, $\mathscr{S}_M$ is a set of subsets of $M$ serving as the range of the set variables, $+_M$ and $\cdot_M$ are binary operations on $M$, $0_M$ and $1_M$ are distinguished elements of $M$, and $<_M$ is a binary relation on $M$. In this book, for a set $x$, $\mathscr{P}(x)$ denotes the set of all subsets of $x$. As always, $\langle \omega, \mathscr{P}(\omega), +, \cdot, 0, 1, < \rangle$ is the standard model of $L_2$.

An $L_2$-formula is *arithmetical* if it contains no set quantifiers, i.e. all of the quantifiers appearing in the formula are number quantifiers. A set of natural numbers is *arithmetical* if it is definable in $\langle \omega, \mathscr{P}(\omega), +, \cdot, 0, 1, < \rangle$ by a $L_2$-formula without set quantifiers. In the following, I inductively define the formula classes $\Sigma_n^0$, $\Pi_n^0$ and $\Delta_n^0$ and the corresponding arithmetical hierarchy.

Let $n$ be a number variable, $t$ be a $L_2$-term not containing $n$ and $\phi$ be a $L_2$-formula. The expressions $\forall n < t, \forall n \leq t, \exists n < t, \exists n \leq t$ are called *bounded* number quantifiers. A *bounded quantifier formula* is a formula $\phi$ such that all of the quantifiers occurring in $\phi$ are bounded number quantifiers. An $L_2$-formula $\phi$ is said to be $\Sigma_1^0$ if it is of the form $\exists m \psi$, where $m$ is a number variable and $\psi$ is a bounded quantifier formula. An $L_2$-formula $\phi$ is said to be $\Pi_1^0$ if $\neg \phi$ is $\Sigma_1^0$. An $L_2$-formula $\phi$ is said to be $\Sigma_{n+1}^0$ if it is of the form $\exists m \psi$, where $m$ is a number variable and $\psi$ is a $\Pi_n^0$ formula. An $L_2$-formula $\phi$ is said to be $\Pi_{n+1}^0$ if $\neg \phi$ is $\Sigma_{n+1}^0$. An $L_2$-formula $\phi$ is $\Delta_n^0$ iff $\phi$ is both $\Sigma_n^0$ and $\Pi_n^0$.

**Definition 1.1** (*Peano Arithmetic*) The system PA is the first-order theory of arithmetic with $L(\mathsf{PA}) = \{0, S, +, \cdot\}$. The system PA consists of the axioms for first-order logic and the following axioms for arithmetic:

(1) $\forall x \forall y (Sx = Sy \rightarrow x = y)$;
(2) $\forall x (Sx \neq 0)$;
(3) $\forall x \forall y (x + 0 = x)$;
(4) $\forall x \forall y (x + Sy = S(x + y))$;
(5) $\forall x (x \cdot 0 = 0)$;
(6) $\forall x \forall y (x \cdot Sy = x \cdot y + x)$;
(7) The scheme of induction: $(\phi(0) \wedge \forall x (\phi(x) \rightarrow \phi(Sx))) \rightarrow \forall x \, \phi(x)$, where $\phi$ is a formula in $L(\mathsf{PA})$ with at least one free variable $x$.

**Definition 1.2** (*Robinson's* Q) The system Q is the sub-theory of PA with axioms for arithmetic consisting of axioms (1)–(6) in Definition 1.1 plus the following axiom: $\forall x (x \neq 0 \rightarrow \exists y (x = Sy))$.

In this book, let $\mathfrak{N} = (\omega, 0, S, +, \cdot)$ denote the standard model of PA where $S$ is the successor function on $\omega$. We say that $\phi \in L(\mathsf{PA})$ is a *true sentence* of arithmetic if $\mathfrak{N} \models \phi$. The system $\mathsf{I}\Sigma_n^0$ is the fragment of PA obtained by restricting the axiom scheme of induction to $\Sigma_n^0$-formulas (see [13]).

For $n \in \omega$, let $\bar{n}$ denote the corresponding numeral for $n$ in $L(\mathsf{PA})$. An $n$-ary relation $R(x_1, \ldots, x_n)$ on $\omega^n$ is *representable* in a theory $T$ if there is a formula

$\phi(x_1, \ldots, x_n)$ in $L(T)$ such that if $R(m_1, \ldots, m_n)$ holds, then $T \vdash \phi(\overline{m_1}, \ldots, \overline{m_n})$; and if $R(m_1, \ldots, m_n)$ does not hold, then $T \vdash \neg\phi(\overline{m_1}, \ldots, \overline{m_n})$. A theory $T$ is said to be *$\omega$-consistent* if there is no formula $\varphi(x)$ such that $T \vdash \exists x \varphi(x)$ and for any $n \in \omega$, $T \vdash \neg\varphi(\overline{n})$. A theory $T$ is *1-consistent* if there is no such a $\Delta_1^0$-formula $\varphi(x)$.

The notion of *interpretation* provides us with a method for comparing different theories in different languages as follows. Let $T$ be a theory in the language $L(T)$, and $S$ a theory in the language $L(S)$. In its most simple form, a translation $I$ of the language $L(T)$ into the language $L(S)$ is specified by the following items.

(1) An $L(S)$-formula $\delta_I(x)$ denoting the domain of $I$.
(2) For each relation symbol $R$ of $L(T)$, an $L(S)$-formula $R_I$ of the same arity.
(3) For each function symbol $F$ of $L(T)$ of arity $k$, an $L(S)$-formula $F_I$ of arity $k + 1$.

The translation $I$ of $L(T)$ into $L(S)$ is called an *interpretation* of $T$ in $S$ if for each function symbol $F$ of $L(T)$ of arity $k$, the formula expressing that $F_I$ is total on $\delta_I$ as follows:

$$(\forall x_0, \ldots \forall x_{k-1}(\delta_I(x_0) \land \cdots \land \delta_I(x_{k-1}) \to \exists y(\delta_I(y) \land F_I(x_0, \ldots, x_{k-1}, y))))$$

is provable in $S$, and the $I$-translations of all axioms of $T$ are provable in $S$.[2]

A theory $T$ is *interpretable* in a theory $S$, denoted by $T \trianglelefteq S$, if there exists an interpretation of $T$ in $S$. If $T$ is interpretable in $S$, then all sentences provable (refutable) in $T$ are mapped, by the interpretation function, to sentences provable (refutable) in $S$. Let $T \lhd S$ denote that $T \trianglelefteq S$ but $S$ is not interpretable in $T$. We express that $T$ and $S$ are mutually interpretable if $T \trianglelefteq S$ and $S \trianglelefteq T$. Interpretability is one among many measures of strength of theories. If $S \lhd T$, then $S$ can be considered weaker than $T$ w.r.t. interpretation; if $S$ and $T$ are mutually interpretable, then $T$ and $S$ are equally strong w.r.t. interpretation. Whenever we express that $S$ is weaker than $T$ w.r.t. interpretation, this just means that $S \lhd T$.

Gödel proved his incompleteness theorems in [16] for a certain formal system P related to Russell-Whitehead's Principia Mathematica and based on the simple theory of types over the natural number series and the Dedekind-Peano axioms (see [9], p. 3). Gödel's original first incompleteness theorem ([16, Theorem VI]) states that for a formal theory $T$ formulated in the language of P and obtained by adding a primitive recursive set of axioms to the system P, if $T$ is $\omega$-consistent, then $T$ is incomplete. Rosser improved Gödel's result by only assuming $T$ to be consistent. The following theorem is a modern reformulation of the Gödel-Rosser incompleteness theorem.

**Theorem 1.1** (Gödel-Rosser incompleteness theorem (G1)) *If $T$ is a recursively axiomatized consistent extension of* PA, *then $T$ is incomplete.*

---

[2]The simplified picture of translations and interpretations above actually describes only one-dimensional, parameter-free, and one-piece translations. For the precise definitions of a multi-dimensional interpretation, an interpretation with parameters and a piece-wise interpretation, I refer to [12, 14, 15] for more details.

The goal of the aforementioned Hilbert's program is to formalize all mathematical statements in a certain logical system and prove the completeness, i.e. all true mathematical statements can be proved in this formalism. It is generally agreed that G1 implies that it is not possible to formalize all of mathematics within a consistent and complete formal system, as any attempt at such a formalism will omit some true mathematical statements. In short, there is no complete recursively axiomatized consistent extension of PA.

In the following, I provide a sketch of the main idea of Gödel's proof of G1. Let $T$ be a recursively axiomatized consistent extension of PA. The three main ideas in Gödel's proof of G1 are the arithmetization of the syntax of $T$, the representability of recursive functions in PA, and Gödel's self-reference/fixed point construction.

Firstly, Gödel gives a recursive arithmetization of the axioms of $T$ (i.e. under Gödel's arithmetization, the set of Gödel's number of axioms of $T$ is recursive). Then we can define certain relations on natural numbers which express certain metamathematical properties of $T$. For instance, one defines a 'proof' relation on $\omega^2$ as follows: $\mathsf{Prf}_T(m, n)$ iff $n$ is the Gödel's number of a proof in $T$ of the formula with Gödel number $m$. Moreover, we can prove that the relation $\mathsf{Prf}_T(m, n)$ is recursive. Secondly, Gödel proves that every recursive relation is representable in PA and hence there is a formula $\phi(x, y)$ which represents $\mathsf{Prf}_T(m, n)$ in PA.

From the representation formula $\phi(x, y)$, we can naturally define the provability predicate $\mathsf{Pr}_T(x)$ as follows: $\mathsf{Pr}_T(x) = \exists y\, \phi(x, y)$. In this book, I use $\ulcorner \phi \urcorner$ to denote the numeral of the Gödel number of formula $\phi$ in $L(\mathsf{PA})$. Thirdly, Gödel constructs the so-called Gödel sentence G which intuitively asserts its own unprovability in $T$, i.e. $T \vdash \mathsf{G} \leftrightarrow \neg\mathsf{Pr}_T(\ulcorner \mathsf{G} \urcorner)$. Finally, Gödel shows that if $T$ is consistent, then G is not provable in $T$; and if $T$ is $\omega$-consistent, then $\neg$G is not provable in $T$.

As essential part of Gödel's above proof is that the provability predicate $\mathsf{Pr}_T(x)$ satisfies the following conditions:

**D1** If $T \vdash \varphi$, then $T \vdash \mathsf{Pr}_T(\ulcorner \varphi \urcorner)$;
**D2** $T \vdash \mathsf{Pr}_T(\ulcorner \varphi \urcorner) \to (\mathsf{Pr}_T(\ulcorner \varphi \to \psi \urcorner) \to \mathsf{Pr}_T(\ulcorner \psi \urcorner))$;
**D3** $T \vdash \mathsf{Pr}_T(\ulcorner \varphi \urcorner) \to \mathsf{Pr}_T(\ulcorner \mathsf{Pr}_T(\ulcorner \varphi \urcorner) \urcorner)$.

The items D1-D3 are called the *Derivability Conditions*.

Next, Gödel announced the second incompleteness theorem in an abstract published in October 1930. Therefore, no consistency proof of systems such as Principia, Zermelo-Fraenkel set theory, or the systems investigated by Ackermann and von Neumann is possible by methods which can be formulated in these systems (see [17], p. 431).

For the proof of Gödel's second incompleteness theorem, we first define the sentence $\mathsf{Con}(T)$ which expresses the consistency of $T$ in $L(\mathsf{PA})$ as follows: $\mathsf{Con}(T) \triangleq \neg\mathsf{Pr}_T(\ulcorner 0 = \bar{1} \urcorner)$. We call $\mathsf{Con}(T)$ the *consistency statement* of $T$. The following is a modern reformulation of Gödel's second incompleteness theorem:

**Theorem 1.2** (Gödel's second incompleteness theorem (G2)) *Let $T$ be a recursively axiomatized extension of PA. If $T$ is consistent, then $\mathsf{Con}(T)$ is not provable in $T$.*

As to a proof of G2, from conditions D1-D3, one can show that $T \vdash \text{Con}(T) \leftrightarrow$ G. Thus, G2 holds: if $T$ is consistent, then $T \nvdash \text{Con}(T)$ since $T \nvdash$ G.[3] For more details of Gödel's proof of G1 and G2, I refer to Chap. 2 of [3].

At the most fundamental level, G1 reveals the difference between the notion of provability in PA and the notion of truth in the standard model of PA. Let Prov be the set of Gödel number of $\phi$ in $L(\text{PA})$ such that $\text{PA} \vdash \phi$ and Truth be the set of Gödel number of $\phi$ in $L(\text{PA})$ such that $\mathfrak{N} \models \phi$. Tarski proved that Truth is not definable in $\mathfrak{N}$. Then G1 reveals the essential difference between Truth and Prov: Prov $\subsetneq$ Truth, i.e. there is a true sentence of arithmetic which is independent of PA. Moreover, Truth is not definable in the standard model $\mathfrak{N}$ but Prov is definable in $\mathfrak{N}$; Truth is not arithmetical but Prov is recursive enumerable; Truth and Prov both are not recursive and not representable in PA. For details of the properties of Truth and Prov, I refer to [3, 18].

**Definition 1.3** Let $T$ be a theory and $\Gamma$ be a class of arithmetical formulas.

(1) $T$ is $\Sigma_n^0$-*definable* if there is a $\Sigma_n^0$ formula $\alpha(x)$ such that $m$ is the Gödel number of some sentence of $T$ iff $\mathfrak{N} \models \alpha(\overline{m})$.
(2) $T$ is $\Sigma_n^0$-*sound* if for all $\Sigma_n^0$ sentences $\phi$, if $T \vdash \phi$, then $\mathfrak{N} \models \phi$.
(3) $T$ is $\Gamma$-*decisive* if for all $\Gamma$ sentences $\phi$, either $T \vdash \phi$ or $T \vdash \neg\phi$ holds.

From G1 and so-called Craig's trick ([4], p. 10), it follows that if a theory $T$ is $\Sigma_1^0$-definable consistent extension of PA, then $T$ is not $\Pi_1^0$-decisive. In [19], Kikuchi and Kurahashi generalized G1 to arithmetically definable theories.

**Theorem 1.3** ([19], Theorem 4.8) *If $T$ is $\Sigma_{n+1}^0$-definable and $\Sigma_n^0$-sound extension of* PA, *then $T$ is not $\Pi_{n+1}^0$-decisive.*[4]

In the following, I examine the generalization of G1 to subsystems of PA. Of course, G1 can be generalized via the notion of interpretation: there is a weak subsystem $T$ of PA such that for any recursively axiomatizable consistent theory $S$, if $T$ is interpretable in $S$, then $S$ is incomplete. Now, I define a general notion of "G1 holds for $T$".

**Definition 1.4** We express that G1 *holds for $T$* if for any recursively axiomatizable consistent theory $S$, if $T$ is interpretable in $S$, then $S$ is incomplete.

**Definition 1.5** Let R be the system consisting of schemes R1-R5 with $L(\text{R}) = \{0, \text{S}, +, \cdot, \leq\}$ where $m, n \in \omega$ and $\overline{n} = S^n(0)$.

R1  $\overline{m} + \overline{n} = \overline{m+n}$;
R2  $\overline{m} \cdot \overline{n} = \overline{m \cdot n}$;
R3  $\overline{m} \neq \overline{n}$  if $m \neq n$;

---

[3]It is not enough to show that $\neg\text{Con}(T)$ is not provable in $T$ only assuming $T$ is consistent. But we could prove that $\text{Con}(T)$ is independent of $T$ by assuming that $T$ is 1-consistent.

[4]The optimality of this generalization is shown by Salehi and Seraji in [20]: there exists a $\Sigma_{n+1}^0$-definable, $\Sigma_{n-1}^0$-sound ($n \geq 1$) and complete theory which contains Q (see Theorem 2.6 in [20]).

R4 $\forall x(x \leq \bar{n} \rightarrow x = \bar{0} \vee \cdots \vee x = \bar{n})$;
R5 $\forall x(x \leq \bar{n} \vee \bar{n} \leq x)$.

The system R is a sub-theory of Q, but while Q is finitely axiomatizable, R is not finitely axiomatizable. The system Q is essentially incomplete and in fact minimal essentially incomplete (see [3], p. 260). By contrast, R is essentially incomplete but it is not minimal essentially incomplete (see [21]).

It is well-known that G1 holds for Robinson's Arithmetic Q (see [18]). Vaught [22] essentially showed that G1 holds for R. A natural question is now: can we find a theory $S$ such that G1 holds for $S$ and $S \lhd R$ (i.e. $S$ is weaker than R w.r.t. interpretation)? Recent progress on this problem shows that R is not the weakest theory w.r.t. interpretation for which G1 holds. In fact, we can find many theories $S$ such that G1 holds for $S$ and $S$ is weaker than R w.r.t. interpretation.[5]

We can also reformulate a general version of G2 based on interpretation.

**Theorem 1.4** (A general version of G2, Visser [14]) *There is no r.e. theory $T$ such that* $Q + Con(T)$ *is interpretable in $T$, i.e.* $Q + Con(T) \ntrianglelefteq T$.

Let $T$ be a recursively axiomatized consistent extension of PA. We express that G2 *holds for $T$* if the consistency statement of $T$ is not provable in $T$. Firstly, whether G2 holds for $T$ depends on the definition of the provability predicate in the consistency statement. From Gödel's proof predicate $\mathsf{Prf}_T(x, y)$, we can define the Rosser provability predicate $\mathsf{Pr}_T^R(x)$ as the formula $\exists y(\mathsf{Prf}_T(x, y) \wedge \forall z \leq y \neg \mathsf{Prf}_T(\dot{\neg}(x), z))$, where $\dot{\neg}$ is a function symbol expressing a primitive recursive function calculating the code of $\neg \phi$ from the code of $\phi$. However, $\mathsf{Con}^R(T) \triangleq \neg \mathsf{Pr}_T^R(\ulcorner 0 = \bar{1} \urcorner)$ *is* provable in $T$ by [24, Proposition 2.1].

**Definition 1.6** Let $T$ be any recursively axiomatized consistent extension of PA and let $\alpha(x)$ be a formula in $L(T)$.

(1) Let formula $\mathsf{Prf}_\alpha(x, y)$ express that $y$ is the Gödel number of a proof in $T$ of the formula with Gödel number $x$ from the set of all sentences satisfying $\alpha(x)$.
(2) Define the provability predicate $\mathsf{Pr}_\alpha(x)$ of $\alpha(x)$ as the formula $\exists y \mathsf{Prf}_\alpha(x, y)$ and the consistency statement $\mathsf{Con}_\alpha(T)$ as the formula $\neg \mathsf{Pr}_\alpha(\ulcorner 0 = \bar{1} \urcorner)$.
(3) We express that $\alpha(x)$ *is a numeration of $T$* if for any $n$, $\mathsf{PA} \vdash \alpha(\bar{n})$ if and only if $n$ is the Gödel number of some $\phi \in T$.
(4) We express that $\alpha(x)$ *is a $\Sigma_n^0$ ($\Pi_n^0$ or $\Delta_n^0$) numeration of $T$* if $\alpha(x)$ is a numeration of $T$ and $\alpha(x)$ is a $\Sigma_n^0$ ($\Pi_n^0$ or $\Delta_n^0$) formula.

Now, I give a new reformulation of G2 via the notion of numeration.

**Theorem 1.5** *Let $T$ be any recursively enumerable consistent extension of PA. If $\alpha(x)$ is any $\Sigma_1^0$ numeration of $T$, then $T \nvdash \mathsf{Con}_\alpha(T)$.*

---

[5]We say $\langle S, T \rangle$ is a recursively inseparable pair if $S$ and $T$ are disjoint r.e. sets of natural numbers, and there is no recursive set $X \subseteq \mathbb{N}$ such that $S \subseteq X$ and $X \cap T = \emptyset$. Cheng [23] shows that for any recursively inseparable pair $\langle A, B \rangle$, there is a theory $U_{\langle A, B \rangle}$ such that G1 holds for $U_{\langle A, B \rangle}$ and $U_{\langle A, B \rangle} \lhd R$.

Clearly, the previous theorem expresses a kind of intensionality of G2: whether G2 as in Theorem 1.5 holds for PA depends on the numeration of PA. Indeed, G2 holds for $\Sigma_1^0$ numerations of PA, but fails for some $\Pi_1^0$ numerations of PA. For example, Feferman [25] constructs a $\Pi_1^0$ numeration $\pi(x)$ of PA such that G2 fails: $\mathrm{Con}_\pi(\mathsf{PA}) \triangleq \neg\mathrm{Pr}_\pi(\ulcorner 0 = \overline{1}\urcorner)$ is provable in PA. Here, I only mention the intensionality problem of G2; I refer to Detlefsen [26], Visser [12, 14], Feferman [25] and Franks [27] and others for more details on the intensionality of G2.

It is generally believed that G1 and G2 show that Hilbert's program for the foundations of mathematics is impossible. Nonetheless, there are dissidents: Detlefsen [28] and more recently Artemov [29]. Moreover, Hilbert's program has inspired various investigations in foundations of mathematics, and there is a *partial* realization. The program of *Reverse Mathematics*, founded by Friedman and developed by Simpson, yields such a partial realization of Hilbert's original program. Now I discuss this program in some detail in the following section.

### 1.1.2  Basics of Reverse Mathematics

In this section, I give a brief introduction to Reverse Mathematics and review some notions and facts used in this book. Reverse Mathematics is a program in foundations of mathematics initiated by Friedman [30, 31] and later developed extensively by Simpson [32] and others. Stillwell provides an excellent introduction for the mathematician-in-the-street in [33]. Simpson writes the following concerning the foundations of mathematics.

> foundations of mathematics is a subject of the greatest mathematical and philosophical importance. Beyond this, foundations of mathematics is a rich subject with a long history, going back to Aristotle and Euclid and continuing in the hands of outstanding modern figures such as Descartes, Cauchy, Weierstraß, Dedekind, Peano, Frege, Russell, Cantor, Hilbert, Brouwer, Weyl, von Neumann, Skolem, Tarski, Heyting, and Gödel (see [32] preface xiii). An excellent reference for the modern era in foundations of mathematics is van Heijenoort [34].

Reverse Mathematics is a highly developed research program whose purpose is to investigate the role of strong set existence axioms in ordinary mathematics. Since Gödel's incompleteness theorem shows that not all of classical mathematics can be reduced to and justified by finitistic mathematics, it is a natural question how much of classical mathematics can be so reduced?

Reverse Mathematics seeks to give a precise answer to this question by investigating which set existence axioms are needed in order to prove theorems of ordinary non-set-theoretic mathematics (see [32]). Classical Reverse Mathematics as developed by Friedman and Simpson takes place in $L_2$ (the language of second-order arithmetic) which is the weakest language that is rich enough to express and develop the bulk of core mathematics, according to Simpson (see [32], p. xiv). I refer to [32, 33] for a comprehensive introduction to Classical Reverse Mathematics. All definitions on SOA in this book are from the standard textbook [32].

Recall that in Sect. 1.1.1 I have defined the notion of $\Sigma_n^0$, $\Pi_n^0$ and $\Delta_n^0$ formulas as part of the arithmetic hierarchy. Now, a set of reals is *analytical* if it is definable in $\langle \omega, \mathscr{P}(\omega), +, \cdot, 0, 1, < \rangle$ by a $L_2$-formula. In the following, I define the notion of $\Sigma_n^1$, $\Pi_n^1$ and $\Delta_n^1$ formulas as part of the analytical hierarchy.

An $L_2$-formula $\phi$ is $\Sigma_1^1$ if it is of the form $\exists X \psi$, where $X$ is a set variable and $\psi$ is an arithmetical formula. An $L_2$-formula $\phi$ is $\Pi_1^1$ if $\neg \phi$ is $\Sigma_1^1$. An $L_2$-formula $\phi$ is $\Sigma_{n+1}^1$ if it is of the form $\exists X \psi$, where $X$ is a set variable and $\psi$ is a $\Pi_n^1$ formula. An $L_2$-formula $\phi$ is $\Pi_{n+1}^1$ if $\neg \phi$ is $\Sigma_{n+1}^1$. An $L_2$-formula $\phi$ is $\Delta_n^1$ if $\phi$ is both $\Sigma_n^1$ and $\Pi_n^1$.

**Definition 1.7** ([32]) The axioms of SOA consist of the universal closures of the following $L_2$-formulas:
(i) Basic axioms:

$$n + 1 \neq 0; \quad m + 1 = n + 1 \rightarrow m = n;$$

$$m + 0 = m; \quad m + (n + 1) = (m + n) + 1;$$

$$m \cdot 0 = 0; \quad m \cdot (n + 1) = (m \cdot n) + m;$$

$$\neg m < 0; \quad m < n + 1 \leftrightarrow (m < n \vee m = n).$$

(ii) Induction axiom: $(0 \in X \wedge \forall n (n \in X \rightarrow n + 1 \in X)) \rightarrow \forall n (n \in X)$.
(iii) Comprehension scheme: $\exists X \forall n (n \in X \leftrightarrow \varphi(n))$, where $\varphi(n)$ is any $L_2$-formula in which $X$ does not occur freely.

In the search for the minimal set existence axioms as part of Reverse Mathematics, the following axiom systems recur constantly: RCA$_0$ (Recursive Comprehension), WKL$_0$ (Weak Konig's Lemma), ACA$_0$ (Arithmetical Comprehension), ATR$_0$ (Arithmetic Transfinite Recursion) and $\Pi_1^1$-CA$_0$ ($\Pi_1^1$-Comprehension). These are the most famous five subsystems of SOA, and are called the 'Big Five'.

**Definition 1.8** (*The Big Five systems*)

(1) RCA$_0$ is the subsystem of SOA consisting of the basic axioms in Definition 1.7(i), the induction axiom in Definition 1.7(ii) restricted to $\Sigma_1^0$ $L_2$-formulas, and the comprehension scheme in Definition 1.7(iii) restricted to $\Delta_1^0$ $L_2$-formulas.
(2) The so-called *Weak König's lemma* is the following statement: every infinite subtree of $2^{<\omega}$ has an infinite path.
(3) WKL$_0$ is defined to be the subsystem of SOA consisting of RCA$_0$ plus the Weak König's lemma.
(4) The *arithmetical comprehension scheme* is the restriction of the comprehension scheme in Definition 1.7(iii) to arithmetical formulas.
(5) ACA$_0$ is the subsystem of SOA whose axioms are the arithmetical comprehension scheme, the basic axioms in Definition 1.7(i) and the induction axiom in Definition 1.7(ii).
(6) ATR$_0$ is a subsystem of SOA consisting of ACA$_0$ plus *arithmetical transfinite recursion* which informally says that the Turing jump operator can be iterated along any countable well-ordering starting at any set.

(7) $\Pi_1^1$-$\mathsf{CA}_0$ is the subsystem of $\mathsf{SOA}$ whose axioms are the basic axioms in Definition 1.7(i), the induction axiom in Definition 1.7(ii), and the comprehension scheme in Definition 1.7(iii) restricted to $\Pi_1^1$ $L_2$-formulas.

Similar to the final item in Definition 1.8, one defines $\Pi_k^1$-$\mathsf{CA}_0$ for $k \in \omega$ as comprehension restricted to $\Pi_k^1$-formulas. The Big Five only constitute a very tiny fragment of $\mathsf{SOA}$ and have strictly increasing strength as follows: $\mathsf{RCA}_0$, $\mathsf{WKL}_0$, $\mathsf{ACA}_0$, $\mathsf{ATR}_0$ and $\Pi_1^1$-$\mathsf{CA}_0$. It is a fact that $\mathsf{SOA} = \bigcup_{k \in \omega} \Pi_k^1$-$\mathsf{CA}_0$ (see [32]).

Assuming the base theory $\mathsf{RCA}_0$ of computable mathematics, the aim of Reverse Mathematics is to find the minimal axioms $A$ such that the statement $B$ of ordinary mathematics is provable from $A$ over $\mathsf{RCA}_0$. A surprising fact is: once the minimal axioms $A$ have been found, we almost always have that $B$ is equivalent to $A$ over $\mathsf{RCA}_0$, i.e. not only can we derive the theorem $B$ from the axioms $A$, i.e. the usual way of doing mathematics, we can also derive the axioms $A$ from the theorem $B$, i.e. the 'reverse' way of doing mathematics (see [35]). More surprisingly, for a statement $B$ of ordinary mathematics, in most cases, either $B$ is provable in $\mathsf{RCA}_0$ or $B$ is equivalent to one of the following systems over $\mathsf{RCA}_0$: $\mathsf{WKL}_0$, $\mathsf{ACA}_0$, $\mathsf{ATR}_0$ and $\Pi_1^1$-$\mathsf{CA}_0$; this is called the Big Five phenomenon (see [35]). Exceptions are classified in the so-called Reverse Mathematics zoo [36].

Based on the results of Reverse Mathematics, we can explore the consequences and limits of some doctrines and programs in foundations of mathematics, for example (see [37]):

(1) constructivism or computable analysis (see Aberth [38], Pour-El/Richards [39]);
(2) finitistic reductionism (see Hilbert [40]);
(3) predicativity (see Weyl [41], Kreisel [42], Feferman [43–45]);
(4) predicative reductionism (see Feferman [46], Simpson [47], and Friedman-McAloon-Simpson [48]);
(5) impredicative or $\Pi_1^1$ analysis (see Buchholz et al. [49]).

The Big Five correspond to foundational programs: constructivism, finitistic reductionism, predicativism, predicative reductionism and impredicativism (see [32]).

More recently, Ulrich Kohlenbach has proposed *higher-order* Reverse Mathematics, which extends the analysis of ordinary mathematics in the context of $\mathsf{SOA}$ to the language of higher-order arithmetic. For higher-order Reverse Mathematics, I refer to works by Kohlenbach [50, 51], Normann-Sanders [35, 52].

In contrast to classical Reverse Mathematics, higher-order Reverse Mathematics makes use of the much richer language *of all finite types*: while $\mathsf{SOA}$ is restricted to numbers and sets of numbers, higher-order arithmetic also has sets of sets of numbers, sets of sets of sets of numbers and etc. In particular, the collection of all finite types $\mathsf{T}$ can be defined inductively by: (i) $0 \in \mathsf{T}$ and (ii) If $\sigma, \tau \in \mathsf{T}$ then $(\sigma \to \tau) \in \mathsf{T}$, where $0$ is the type of natural numbers, and $\sigma \to \tau$ is the type of mappings from objects of type $\sigma$ to objects of type $\tau$. For example, $1 \triangleq 0 \to 0$ is the type of functions from numbers to numbers where $n + 1 \triangleq n \to 0$. Viewing sets as given by their characteristic functions, $\mathsf{SOA}$ only includes objects of type $0$ and $1$; and higher-order arithmetic includes objects of higher finite types.

Next, the language of third-order arithmetic (denoted by $Z_3$) has three sorts of variables: variables of type 0 ranging over $\omega$ to be denoted by $x^0, y^0, z^0, \ldots$; variables of type 1 ranging over $\mathscr{P}(\omega)$ to be denoted by $x^1, y^1, z^1, \ldots$; and variables of type 2 ranging over $\mathscr{P}(\mathscr{P}(\omega))$ to be denoted by $x^2, y^2, z^2, \ldots$. Correspondingly, there are three sorts of quantifiers: $\exists^0 x^0, \forall^0 x^0, \exists^1 x^1, \forall^1 x^1$ and $\exists^2 x^2, \forall^2 x^2$ qualifying over the variables of type 0, 1 and 2. The terms of type 0 are the same as for $Z_2$. The atomic formulas of $Z_3$ are the equations between terms of type 0, $x^0 \in x^1$ and $x^1 \in x^2$, where $x^0, x^1, x^2$ are respectively variables of types 0, 1 and 2. Formulas of $Z_3$ are built up as usual using the connectives and different sorts of quantifiers. Axioms of third-order arithmetic include axioms (1)–(6) and the scheme of induction in Definition 1.1, and the following scheme of axioms: (i) $(\exists x^1)(\forall x^0)(x^0 \in x^1 \leftrightarrow \phi)$, where $\phi$ is any formula in $L(Z_3)$ in which $x^1$ is not free; (ii) $(\exists x^2)(\forall x^1)(x^1 \in x^2 \leftrightarrow \phi)$, where $\phi$ is any formula in $L(Z_3)$ in which $x^2$ is not free.

Kohlenbach argues in [50] that there are good reasons to extend the context of classical Reverse Mathematics to the language of arithmetic in all finite types: for example, the availability of variables for arbitrary (not just continuous) functions within the language allows for an extension of Reverse Mathematics (see [50]). The base system of the classical Reverse Mathematics is $\mathsf{RCA}_0$. Kohlenbach introduced the system $\mathsf{RCA}_0^\omega$ as the base system of higher-order Reverse Mathematics which is a conservative finite type extension of $\mathsf{RCA}_0$. For the definition of $\mathsf{RCA}_0^\omega$, I refer to [35, 50]. Note that $\mathsf{RCA}_0^\omega$ and $\mathsf{RCA}_0$ prove the same $L_2$-sentences (see [50], Sect. 2). The language of finite types allows to treat various analytical notions in a more direct way, applies to many more principles and produces interesting extensions of classical Reverse Mathematics (see [50]).

Finally, $\mathfrak{L}_{st}$ denotes the language of set theory: first-order predicate calculus with equality and the binary predicate symbol $\in$. In this book, ZFC denotes Zermelo-Fraenkel Set Theory with the Axiom of Choice in $\mathfrak{L}_{st}$[6]; ZF denotes Zermelo-Fraenkel Set Theory without the Axiom of Choice; $\mathsf{ZFC}^-$ denotes ZFC with the Power Set Axiom deleted and Collection instead of Replacement.[7]

In this book, I work in the set-theoretic definition of higher-order arithmetic via coding defined as follows: $Z_2, Z_3$ and $Z_4$ are the corresponding set-theoretical axiomatic systems for second-order arithmetic, third-order arithmetic and fourth-order arithmetic (similarly, I can define $Z_n$ for $n > 4$).

**Definition 1.9**

(i) $Z_2 = \mathsf{ZFC}^- +$ Every set is countable.
(ii) $Z_3 = \mathsf{ZFC}^- + \mathscr{P}(\omega)$ exists + Every set is of cardinality $\leq \beth_1$.[8]
(iii) $Z_4 = \mathsf{ZFC}^- + \mathscr{P}(\mathscr{P}(\omega))$ exists + Every set is of cardinality $\leq \beth_2$.

---

[6]ZFC consists of the following axioms: Existence, Extensionality, Comprehension, Pairing, Union, Powerset, Foundation, Replacement, Axiom of Choice and Axiom of Infinity.

[7]For a discussion about the proper axiomatic framework for set theory without the power set axiom, I refer to [53].

[8]For $n \in \omega$, $\beth_{n+1}$ is the cardinality of $2^{\beth_n}$ and $\beth_0 = \omega$.

In this book, I use $\alpha, \beta, \gamma, \delta \cdots$ to denote ordinals and $\kappa, \lambda, \mu, \nu \cdots$ to denote infinite cardinals; let $tc(X)$ denote the transitive closure of $X$ (the smallest transitive set $\supseteq X$). For a set $X$, $|X|$ denotes its cardinality, and for infinite cardinals $\kappa$, $H_\kappa :=$ $\{x \mid |tc(x)| < \kappa\}$. Let $\mathsf{HC}$ denote $H_{\omega_1}$, let $\mathsf{CH}$ denote the Continuum Hypothesis, and let $\mathsf{GCH}$ denote the Generalized Continuum Hypothesis. Since "any set is countable" is equivalent to $V = \mathsf{HC}$, $\mathsf{Z}_2 = \mathsf{ZFC}^- + V = \mathsf{HC}$. Under $\mathsf{CH}$, I have that

$$\mathsf{Z}_3 = \mathsf{ZFC}^- + \mathscr{P}(\omega) \text{ exists} + \text{Every set is of cardinality } \leq \omega_1.$$

Under $\mathsf{GCH}$, $\mathsf{Z}_4 = \mathsf{ZFC}^- + \mathscr{P}(\mathscr{P}(\omega))$ exists + Every set is of cardinality $\leq \omega_2$. Also note that $\mathsf{Z}_3 \vdash$ "$H_{\omega_1} \models \mathsf{Z}_2$" and $\mathsf{Z}_4 \vdash$ "$H_{\beth^+} \models \mathsf{Z}_3$".

Finally, I discuss the relationship between $\mathsf{SOA}$ and $\mathsf{Z}_2$. A similar relationship between higher-order arithmetic defined in the language of finite types and the set-theoretic hierarchy of $\mathsf{Z}_n$ can also be established. $\mathsf{SOA}$ is defined in the language $L_2$ and $\mathsf{Z}_2$ is defined in the language $\mathfrak{L}_{st}$. Note that $\langle \omega, \mathscr{P}(\omega), +, \cdot, 0, 1, < \rangle \models \mathsf{SOA}$, $\langle V_{\omega+1}, \in \rangle \models \mathsf{Z}_2$ and $\langle \mathsf{HC}, \in \rangle \models \mathsf{Z}_2$.

**Fact 1.6** (Proposition 1.4, [54]) $\mathsf{ZFC}^-$ *with additionally a definable well-ordering of the universe, is a* $\Pi_4^1$ *conservative extension of* $\mathsf{SOA}$.

**Fact 1.7** (Corollary 1.15, [55]) *Over* $\mathsf{RCA}_0$, *the following are equiconsistent:*

(1) $\mathsf{SOA}$, *i.e. the scheme consisting of the axiom* $\Pi_n^1\text{-}\mathsf{CA}_0$ *for each* $n \in \omega$;
(2) *The scheme consisting of the axiom* $Det(n\text{-}\boldsymbol{\Pi}_3^0)$ *for each* $n \in \omega$[9];
(3) $\mathsf{ZFC}^-$.

**Definition 1.10** Let $M$, $N$ be two structures respectively in the language of $\mathscr{L}_1$ and $\mathscr{L}_2$. We express that $M$ and $N$ are bi-interpretable if there exists a recursive function $\ulcorner\varphi\urcorner \mapsto \ulcorner\varphi^*\urcorner$ such that $M \models \varphi \Leftrightarrow N \models \varphi^*$ where $\varphi$ is a formula in $\mathscr{L}_1$ and $\varphi^*$ in $\mathscr{L}_2$.

**Fact 1.8** (Folklore)

(1) *Structures* $\langle \omega, \mathscr{P}(\omega), +, \cdot, \in \rangle$, $\langle V_{\omega+1}, \in \rangle$ *and* $\langle \mathsf{HC}, \in \rangle$ *are bi-interpretable.*
(2) $\mathsf{SOA}$ *and* $\mathsf{Z}_2$ *are mutually interpretable.*

### 1.1.3 Overview of Incompleteness for Higher-Order Arithmetic

Gödel's proof of his incompleteness theorem makes use of meta-mathematics, while the associated Gödel sentence has a clear meta-mathematical/logical nature, i.e. it is devoid of real mathematical content. In particular, from a purely mathematical point of view, Gödel's sentence is artificial and not mathematically interesting. A

---

[9]For the definition of $Det(n\text{-}\boldsymbol{\Pi}_3^0)$, see Definition 1.28.

natural question is then: can we find true sentences not provable in PA with a natural formulation and interesting mathematical content?

This problem received a lot of attention because despite Gödel's incompleteness theorems, one could still cherish the hope that all natural and mathematically interesting sentences about natural numbers are provable or refutable in PA, and that elementary arithmetic is complete w.r.t. natural and mathematically interesting sentences (whatever those turn out to be). However, many examples have been found of sentences independent of PA with real mathematical content. The survey paper [56] provides a good overview on the state of art up to autumn 2006. In this section, I give an overview of known independent sentences of PA with natural and mathematically interesting contents.

The first striking example of a mathematically natural statement independent of PA was the *Paris-Harrington Principle*, proposed in [57] and which generalizes the *finite Ramsey theorem*. This principle has a clear combinatorial flavor and does not refer to the arithmetization of syntax and provability.

I first introduce the infinite Ramsey Theorem and the finite Ramsey Theorem. For set $X$ and $n \in \omega$, let $[X]^n$ be the set of all $n$-elements subset of $X$. I identify $n$ with $\{0, \ldots, n-1\}$.

**Definition 1.11** (*Infinite Ramsey Theorem*) Let $n$ and $k$ be positive natural numbers. For any function $f : [\omega]^n \to k$, there exists an infinite set $Y \subseteq \omega$ such that $Y$ is homogeneous for $f$, i. e. the function $f$ restricted to $[Y]^n$ is constant.

**Definition 1.12** (*Finite Ramsey's theorem*) $(\forall p, k, n)(\exists N)(\forall F : [N]^p \to k)(\exists Y \subseteq N)[|Y| \geq n \wedge F \restriction [Y]^p$ is constant]

I refer to [58] for more details on the previous definition. Moreover, the least such $N$ depending on $p, k, n$ is denoted by $N(p, k, n)$ and is called the *Ramsey function*. Erdös-Rado [59] show that $N(n, n, n)$ is bounded by a super-exponential function where the iterations of the exponential function depend linearly on $n$ (see [81]).

**Definition 1.13** (*Paris-Harrington Principle* (PH), [57]) For all $m, n, c \in \mathbb{N}$, there is $N \in \mathbb{N}$ such that for all $f : [N]^m \to c$, we have

$$(\exists H \subseteq N)(|H| \geq n \wedge H \text{ is homogeneous for } f \wedge |H| > \min(H)).$$

Paris-Harrington then established the following theorem in [57].

**Theorem 1.9** *The principle* PH *is true but not provable in* PA.

Now, PH is of the form $\forall x \exists y \psi(x, y)$ where $\psi$ is a $\Delta_0^0$ formula. It can be shown that for any given natural number $n$, PA $\vdash \exists y \psi(\bar{n}, y)$, i.e. all particular *instances* of PH are provable in PA.

Following PH, many other mathematically natural statements independent of PA with combinatorial or number-theoretic content were formulated: the Kanamori-McAloon principle [60], the Kirby-Paris sentence [61], the Hercules-Hydra game [61], the Worm principle [62, 63], the flipping principle [64], the arboreal statement

[65], Pudlák's Principle [66, 67], the kiralic and regal principles [68] (see [56], p. 40). In fact, all of these principles are equivalent to PH (see [56], p. 40).

An interesting and amazing fact is: all the above mathematically natural principles are in fact provably equivalent in PA to a certain *meta-mathematical sentence*, as follows. Consider the following reflection principle for $\Sigma_1^0$ sentences: for any $\Sigma_1^0$ sentence $\phi$ in $L(\mathsf{PA})$, if $\phi$ is provable in PA, then $\phi$ is true. Using the arithmetization of syntax, one can write this principle as a sentence of $L(\mathsf{PA})$ and denote it by $\mathsf{Rfn}_{\Sigma_1^0}(\mathsf{PA})$ (see [3, p. 301]). McAloon has shown that $\mathsf{PA} \vdash \mathsf{PH} \leftrightarrow \mathsf{Rfn}_{\Sigma_1^0}(\mathsf{PA})$ (see [3], p. 301), and similar equivalences can be established for the above other independent principles. Equivalently, all these principles are equivalent to so-called 1-consistency of PA (see [9, p. 36], [62, p. 3] and [3, p. 301]). This phenomenon indicates that the difference between mathematical and meta-mathematical statements is not as huge as we might have expected. The above principles are provable in fragments of SOA and are more complex than Gödel's sentence: Gödel's sentence is equivalent to $\mathsf{Con}(\mathsf{PA})$ in PA; but all these principles are not only independent of PA but also independent of $\mathsf{PA} + \mathsf{Con}(\mathsf{PA})$ (see [9, p. 36] and [3, p. 301]).

To make the reader have a better sense of mathematically natural independent statements of PA, I now discuss two such examples equivalent to PH: the Kanamori-McAloon principle in [60] and the Kirby-Paris sentence in [61].

**Definition 1.14**   A $m$-ary function $f$ is called *regressive* if $f(x_0, x_1, \ldots, x_{m-1}) < x_0$ for all $x_0 < x_1 < \cdots < x_{m-1}$. A set $H$ is called *min-homogeneous* if for all $c_0 < c_1 < \cdots < c_{m-1}$ and $c_0 < d_1 < \cdots < d_{m-1}$ in $H$, $f(c_0, c_1, \ldots, c_{m-1}) = f(c_0, d_1, \ldots, d_{m-1})$.

Kanamori and McAloon [69] introduced an important principle KM of combinatorial contents which is independent of PA.

**Definition 1.15** (*Kanamori-McAloon Principle* (KM), [69]) $(\forall m, a, n)(\exists b)$ (for every regressive function $f$ defined on $[a, b]^m$, there is a min-homogeneous set $H \subseteq [a, b]$ of size at least $n$).

**Theorem 1.10**   (Kanamori-McAloon, [69]) *The principle* KM *is true but unprovable in* PA.

**Definition 1.16**   Given $m, n \in \omega$, we define the *base $n$ representation of $m$* as ($n > 1$): first write $m$ as the sum of powers of $n$, then write each exponent as the sum of powers of $n$; repeat with exponents of exponents and so on until the representation stabilizes. We define the number $G_n(m)$ as: if $m = 0$ let $G_n(m) = 0$; otherwise let $G_n(m)$ be the number produced by replacing every $n$ in the base $n$ representation of $m$ by $n + 1$ and then subtracting 1. The *Goodstein sequence* for the number $m$ is now defined recursively as follows: $m_0 = m$ and $m_k = G_{k+1}(m_{k-1})$ for $k > 0$.

As an example, 266 stabilizes at the representation $2^{2^{2+1}} + 2^{2+1} + 2^1$ and $G_2(266) = 3^{3^{3+1}} + 3^{3+1} + 2$. The procedure of constructing the Goodstein sequence $m_k$ can be defined in $L(\mathsf{PA})$ (see [61]).

**Definition 1.17** (*Kirby-Paris sentence*, [61]) The *Kirby-Paris sentence* denotes the following sentence of number-theoretic contents in $L(\mathsf{PA})$: $\forall m \exists k \, (m_k = 0)$.

**Theorem 1.11** (Kirby-Paris, [61]) *The Kirby-Paris sentence is true but not provable in* $\mathsf{PA}$.

Similarly to $\mathsf{PH}$, the Kirby-Paris sentence is a $\Pi_2^0$ sentence independent of $\mathsf{PA}$ but all particular instances are provable in $\mathsf{PA}$.

Incompleteness would now be complete without mentioning the work of Harvey Friedman: a central figure in the research on the foundations of mathematics after Gödel; he has done a lot of research on concrete mathematical incompleteness. The following quote is telltale:

> the long range impact and significance of ongoing investigations in the foundations of mathematics is going to depend greatly on the extent to which the Incompleteness Phenomena touche normal concrete mathematics (see [70], p. 7).

In the following, I give a brief introduction to Friedman's work on concrete mathematical incompleteness. In his early work, Friedman examines how one uses large cardinals in an essential and natural way in number theory, as follows.

> the quest for a simple meaningful finite mathematical theorem that can only be proved by going beyond the usual axioms for mathematics has been a goal in the foundations of mathematics since Gödel's incompleteness theorems (see [71], p. 805).

He showed in [71, 72] that there are many mathematically natural combinatorial statements in $L(\mathsf{PA})$ that are neither provable nor refutable in $\mathsf{ZFC}$ or $\mathsf{ZFC}+$ large cardinals.

Friedman's more recent book [70] is a comprehensive study of concrete mathematical incompleteness. Most of it is devoted to concrete mathematical incompleteness that arises in *Boolean Relation Theory*. In [70], Friedman uses the following working definition of the mathematically concrete statements: mathematical statements concerning Borel measurable sets and functions of finite rank in and between complete separable metric spaces (see [70] p. 53). Friedman also argues that concrete mathematical incompleteness begins at the level of Borel measurable sets and functions on complete separable metric spaces.

Friedman studies concrete mathematical incompleteness over different systems, ranging from weak subsystems of $\mathsf{PA}$ to higher-order arithmetic and $\mathsf{ZFC}$. Friedman lists many concrete mathematical statements in $L(\mathsf{PA})$ that are independent of subsystems of $\mathsf{PA}$ or stronger theories like higher-order arithmetic and set theories. To provide an idea of the contents of [70], I list some of sections dealing with concrete mathematical incompleteness (see [70] for the relevant definitions).

- Section 0.5 on Incompleteness in Exponential Function Arithmetic;
- Section 0.6 on Incompleteness in Primitive Recursive Arithmetic,
- Single Quantifier Arithmetic, $\mathsf{RCA}_0$, and $\mathsf{WKL}_0$;
- Section 0.7 on Incompleteness in Nested Multiply Recursive Arithmetic and Two Quantifier Arithmetic;

- Section 0.8 on Incompleteness in Peano Arithmetic and $\mathsf{ACA_0}$;
- Section 0.9 on Incompleteness in Predicative Analysis and $\mathsf{ATR_0}$;
- Section 0.10 on Incompleteness in Iterated Inductive Definitions and $\Pi_1^1\text{-}\mathsf{CA_0}$;
- Section 0.11 on Incompleteness in second-order Arithmetic and $\mathsf{ZFC}^-$;
- Section 0.12 on Incompleteness in Russell Type Theory and Zermelo Set Theory;
- Section 0.13 on Incompleteness in $\mathsf{ZFC}$ using Borel Functions;
- Section 0.14 on Incompleteness in $\mathsf{ZFC}$ using Discrete Structures.

In the following, I discuss some of Friedman's examples of concrete mathematical theorems not provable in subsystems of $\mathsf{SOA}$ stronger than $\mathsf{PA}$, and examples of concrete mathematical theorems provable in $\mathsf{Z_3}$ but not provable in $\mathsf{Z_2}$.

**Definition 1.18** A *tree* is a partially ordered set with the least element such that the set of all predecessors of every point is linearly ordered.

Now, the Infinite Kruskal's Theorem says that if $\langle T_i : i \in \omega \rangle$ is a countable sequence of finite trees then there are $i < j$ in $\omega$ such that $T_i \prec T_j$, i.e. there is an infimum-preserving embedding from $T_i$ into $T_j$ (see [56, pp. 43–44]). Friedman [73] proves that the following finite version of the Infinite Kruskal's Theorem is not provable in $\mathsf{ATR_0}$, a theory stronger than $\mathsf{PA}$:

**Definition 1.19** (*Friedman's Principle*) For all $k$, there is $N$ such that if $\langle T_i : 1 \leq i \leq N \rangle$ is a sequence of finite trees such that for all $i \leq N$ we have $|T_i| \leq k + i$, then there are $i, j \leq N$ such that $i < j$ and $T_i \prec T_j$.

Next, let $\alpha$ be *Otter's constant* defined as $\alpha = \frac{1}{p}$ where $p$ is the radius of convergence of $\sum_{i=0}^{\infty} t_i z^i$ where $t_i$ is the number of finite trees of size $i$. Weiermann shows in [80] that for any primitive recursive real number $r$, we have

(1) if $r \leq \frac{1}{\log_2(\alpha)}$ then the statement with the condition $|T_i| \leq k + i$ in Friedman's Principle replaced by $|T_i| \leq k + r \cdot \log_2(i)$ is provable in $\mathsf{I}\Sigma_1$;
(2) if $r > \frac{1}{\log_2(\alpha)}$ then it is not provable in $\mathsf{PA}$.

Another example is a theorem by Friedman, Robertson, and Seymour on the unprovability of the Graph Minor Theorem (see [74]).

**Definition 1.20** For graphs $G$, $H$, we say that $H$ is a minor of $G$ if $H$ is obtained from $G$ by a succession of three elementary operations: edge removal, edge contraction and removal of an isolated vertex.

**Definition 1.21** (*Graph Minor Theorem*) For every $k$, there is $N$ such that if $\{G_i\}_{i=1}^{N}$ is a sequence of finite graphs such that for all $i \leq N$ we have $|G_i| \leq k + i$, then for some $i < j \leq N$, $G_i$ is a minor of $G_j$.

The Graph Minor Theorem is not provable in $\Pi_1^1\text{-}\mathsf{CA_0}$, a rather strong subsystem of the second-order arithmetic (see [56], p. 45). The exact logical strength of this theorem is being investigated in [75].

In [70], Friedman provides a number of concrete mathematical statements provable in third-order arithmetic but independent of $\mathsf{SOA}$: Many other examples of concrete mathematical incompleteness and the discussion of this subject in 1970s–1980s can be found in the four volumes [76–79]. I refer to [80]–[85] for more examples and discussions of mathematically independent statements. I refer to [70] for new advances in Boolean Relation Theory and for more examples of concrete mathematical incompleteness. In this book, I will discuss an example of concrete mathematical theorems expressible in SOA but not provable in $Z_3$.

Finally, I discuss what fragments of the axiom of determinacy are provable in fragments of SOA and give examples of concrete mathematical incompleteness over SOA based on determinacy hypotheses. In the following, I first review some definitions and facts about determinacy and descriptive set theory. My basic references about the theory of determinacy are [86, 87].

In this book, $\omega^\omega$ and $\mathbb{R}$ both denote the set of real numbers. A metric space is *separable* if it has a countable dense subset; it is *complete* if every Cauchy sequence converges. A *Polish space* is a topological space that is homeomorphic to a separable complete metric space.

A typical example of Polish space is *Baire space* $\mathcal{N}$, i.e. the space of all infinite sequences of natural numbers, $\langle a_n : n \in \omega \rangle$, with the topology induced from $\langle O(s) = \{f \in \mathcal{N} : s \subseteq f\} : s \in \omega^{<\omega} \rangle$ which form a basis for the topology.

**Definition 1.22** (*Games*) For $A \subseteq \omega^\omega$, $G_A$ is the following two-player game in which player I and player II alternately play natural numbers.

$$G_A : \begin{array}{c|ccccc} \mathrm{I} & n_0 & & n_2 & \cdots & n_{2t} & & \cdots \\ \hline \mathrm{II} & & n_1 & & n_3 & \cdots & n_{2t+1} & \cdots \end{array}$$

Let $x = (n_0, n_1, n_2, \ldots, n_{2t}, n_{2t+1} \cdots) \in \omega^\omega$. The real $x$ is called a round of the game. We say that Player I wins $G_A$ if $x \in A$; otherwise Player II wins $G_A$.

**Definition 1.23** (*Game theoretic notations*)

(i) A *strategy* for Player I is a function $\sigma : \bigcup_{i \in \omega} \omega^{2i} \to \omega$.

(ii) Let $\sigma * y$ be the real produced when Player I follows $\sigma$ and Player II plays $y$. Then $\sigma$ is a *winning strategy* for Player I in $G_A$ iff for all $y \in \omega^\omega$, $\sigma * y \in A$. i.e. Player I always wins $G_A$ by following $\sigma$ no matter how Player II plays. The corresponding notions for Player II are defined similarly.

(iii) Define that $\sigma * y$ is the round in which Player II plays $y$ against $\sigma$ and $x * \tau$ is the play in which Player I plays $x$ against $\tau$ where $\tau$ is a strategy for Player II. Define that $x * y$ is the resulting real in a play in which Player I plays $x$ and Player II plays $y$. In this case we let $(x * y)_I = x$ and $(x * y)_{II} = y$.[10] If $\sigma$ is a strategy for Player I and $\tau$ is a strategy for Player II, we write $\sigma * \tau$ for the real produced by playing the strategies against one another.

(iv) Given a real $x$, if $\sigma$ is a winning strategy in $G_A$ for player I, we say that $x$ is *consistent* with $\sigma$ if $x = \sigma * y$ for some $y \subseteq \omega$. Similarly, if $\tau$ is a winning strategy in $G_A$ for player II, we say that $x$ is consistent with $\tau$ if $x = y * \tau$ for some $y \subseteq \omega$.

(v) The set $A$ is *determined*, denoted $Det(A)$, if one of the players has a winning strategy in the game $G_A$.

---

[10] $(\sigma * y)_I$ is the real Player I plays in a play in which Player I follows the strategy $\sigma$ against II's play of $y$ (similarly for $(x * \tau)_{II}$).

**Definition 1.24** ([86, 87]) For each $\alpha < \omega_1$, we define the collections $\Sigma^0{}_\alpha$ and $\Pi^0{}_\alpha$ of subsets of Baire space $\omega^\omega$ as follows:

(1) $\Sigma^0_1$ = the collection of all open sets;
(2) $\Pi^0_1$ = the collection of all closed sets;
(3) $\Sigma^0_\alpha$ = the collection of all $A = \bigcup_{n=0}^{\infty} A_n$, where each $A_n$ belongs to $\Pi^0_\beta$ for some $\beta < \alpha$;
(4) $\Pi^0_\alpha$ = the collection of all complements of sets in $\Sigma^0_\alpha$;
(5) Borel = $\bigcup_{\alpha < \omega_1} \Sigma^0_\alpha$.

A set $A \subseteq \mathcal{N}$ is called *analytic* if there exists a continuous function $f : \mathcal{N} \to \mathcal{N}$ such that $A = ran(f)$. The *projection* of a set $S \subseteq \mathcal{N} \times \mathcal{N}$ (into $\mathcal{N}$) is the set $P = \{x \in \mathcal{N} : \exists y \, (x, y) \in S\}$.

**Definition 1.25** ([86, 87]) For each $n \geq 1$, we define the collections $\Sigma^1_n, \Pi^1_n$ and $\Delta^1_n$ of subsets of $\mathcal{N}$ as follows:

(1) $\Sigma^1_1$ = the collection of all analytic sets;
(2) $\Pi^1_1$ = the complements of analytic sets;
(3) $\Sigma^1_{n+1}$ = the collection of the projections of all $\Pi^1_n$ sets in $\mathcal{N} \times \mathcal{N}$;
(4) $\Pi^1_n$ = the complements of the $\Sigma^1_n$ sets in $\mathcal{N}$;
(5) $\Delta^1_n = \Sigma^1_n \cap \Pi^1_n$.

The sets belonging to resp. $\Sigma^1_n, \Pi^1_n$ and $\Delta^1_n$ are called *projective* sets.

We can reformulate the hierarchy of projective sets in terms of the lightface hierarchy $\Sigma^1_n, \Pi^1_n$ and $\Delta^1_n$ and its relativization for real parameters.

**Definition 1.26** ([86, 87])

(1) Given $a \in \mathcal{N}$, $A \subseteq \mathcal{N}$ is $\Sigma^1_1(a)$, *i.e.* $\Sigma^1_1$ in $a$, if there exists a set $D$ recursive in $a$ such that for all $x \in \mathcal{N}$, $x \in A$ iff $(\exists y \in \omega^\omega)(\forall n \in \omega) \, D(x \restriction n, y \restriction n, a \restriction n)$;
(2) $A \subseteq \mathcal{N}$ is $\Pi^1_n$ (in $a$) if the complement of $A$ is $\Sigma^1_n$ (in $a$);
(3) $A \subseteq \mathcal{N}$ is $\Sigma^1_{n+1}$ (in $a$) if it is the projection of a $\Pi^1_n$ (in $a$) subset of $\mathcal{N} \times \mathcal{N}$;
(4) $A \subseteq \mathcal{N}$ is $\Delta^1_n$ (in $a$) if it is both $\Sigma^1_n$ and $\Pi^1_n$ (in $a$).

**Fact 1.12** ([87], Prop. 12.6) *Suppose that $A \subseteq \mathcal{N}$ and $n > 0$. Then:*

*(1)  $A \in \Sigma^0_n$ if and only if $A \in \Sigma^0_n(a)$ for some $a \in \omega^\omega$; similarly for $\Pi^0_n$.*
*(2)  $A \in \Sigma^1_n$ if and only if $A \in \Sigma^1_n(a)$ for some $a \in \omega^\omega$; similarly for $\Pi^1_n$.*

**Definition 1.27** A *pointclass* $\Gamma$ is a collection of sets such that each $P$ in $\Gamma$ is a subset of Baire space $\mathcal{N}$. Given a pointclass $\Gamma$, $Det(\Gamma)$ is the statement that any set of reals in $\Gamma$ is determined.

Let $\leq_T$ and $\equiv_T$ denote, respectively, Turing reducibility and Turing equivalence. Given $A \subseteq \omega^\omega$, $A$ is Turing closed if for any $x \in A$ and $y \in \omega^\omega$, if $x \equiv_T y$, then $y \in A$. $Det(Turing\text{-}\Gamma)$ is the statement that if $X \subseteq \omega^\omega$ is in $\Gamma$ and Turing closed, then $X$ is determined. Note that $Det(\Gamma)$ implies $Det(Turing\text{-}\Gamma)$. Finally, PD denotes the statement that all projective sets of reals are determined.

Friedman [88] shows that Borel Determinacy is not provable in ZFC minus the power set axiom. Moreover, Borel Determinacy is not provable using only countably many transfinite iterations of the power set operation: one needs $\omega_1$ many iterations of the power set operation to prove it. Martin [89] showed that Borel Determinacy is provable in ZFC and provided a level by level analysis of Borel hierarchy and the number of iterations of the power set needed to establish determinacy at those levels. Friedman [88] showed that $Det(\Sigma_5^0)$ is not provable in SOA and Martin [90] improved this to $Det(\Sigma_4^0)$. Finally, Montalbán and Shore [54] established the precise bounds for the amount of determinacy provable in SOA.

**Definition 1.28** A set of reals $A$ is $n$-$\Pi_3^0$ if there are $\Pi_3^0$ sets $A_0, A_1 \cdots, A_n = \emptyset$ such that $x \in A \Leftrightarrow$ the least $i \in \omega$ s.t. $x \notin A_i$ is odd. We say that the sequence $\{A_m : m \le n\}$ represents $A$ as a $n$-$\Pi_3^0$ set.[11]

We use a similar definition for the lightface case. Let $\omega$-$\Pi_3^0 = \bigcup_{n \in \omega} n$-$\Pi_3^0$, all finite boolean combinations of $\Pi_3^0$ sets.

**Theorem 1.13** ([54], Theorems 1.1 and 1.2)

*(1) For each $n \ge 1$, $\Pi_{n+2}^1$-CA$_0 \vdash Det(n$-$\Pi_3^0)$.*
*(2) For every $n \ge 1$, $\Delta_{n+2}^1$-CA does not prove $Det(n$-$\Pi_3^0)$.*

As a corollary of Theorem 1.13, we have the following facts.

(1) $Det(n$-$\Pi_3^0)$ is provable in SOA for any $n \ge 1$;
(2) $Det(\omega$-$\Pi_3^0)$ is not provable in SOA;
(3) $Det(\Delta_4^0)$ is not provable in SOA.

In fact, there is a natural mathematical $\Sigma_2^1$ formula $\varphi(x)$ with only one free variable saying that specific games have strategies and containing no references to provability, such that SOA $\nvdash \forall m\, \varphi(m)$ and for any $n \in \omega$, SOA $\vdash \varphi(\overline{n})$ (see [54, Theorem 1.6]). Moreover, Montalbán and Shore [55] prove that $Det(\omega$-$\Pi_3^0)$ implies the consistency of SOA and more.

### 1.1.4 Basics of Set Theory

In this section, I review some notions and facts in Set Theory used in this book. My definitions and notations are standard. I refer to standard textbooks as [86, 87, 91–94] for the definitions and notations I use. Basic Set Theory (BS) consists of the following schema of axioms in $\mathfrak{L}_{st}$ [94]:

**Definition 1.29** (*Basic Set Theory*)

(1) Extensionality: $\forall x \forall y [\forall z (z \in x \leftrightarrow z \in y) \rightarrow (x = y)]$.
(2) Pairing: $\forall x \forall y \exists z \forall v [(v \in z) \leftrightarrow (v = x \lor v = y)]$.

---

[11] We may assume that $A_i$ are descending. i.e. $A_i \supseteq A_{i+1}$ by replacing $A_i$ by $\bigcap_{j \le i} A_j$.

(3)  Union: $\forall x \exists y \forall z[(z \in y) \leftrightarrow (\exists u \in x)(z \in u)]$.
(4)  Infinity: $\exists x[\mathsf{Ord}(x) \land (x \neq 0) \land (\forall y \in x)(\exists z \in x)(y \in z)]$.
(5)  Cartesian Product: $\forall x \forall y \exists z \forall v[v \in z \leftrightarrow (\exists a \in x)(\exists b \in y)(v = (a, b))]$.
(6)  Induction Schema: $\forall \mathbf{a}[\forall x((\forall y \in x)\Phi(y, \mathbf{a}) \to \Phi(x, \mathbf{a})) \to \forall x \Phi(x, \mathbf{a})]$, where $\Phi$ is any formula of $\mathfrak{L}_{st}$ with free variables among $x, \mathbf{a}$.
(7)  $\Sigma_0$ comprehension schema: $\forall \mathbf{a} \forall x \exists y \forall z[(z \in y) \leftrightarrow (z \in x \land \Phi(z, \mathbf{a}))]$, where $\Phi(x, \mathbf{a})$ is a $\Sigma_0$ formula in $\mathfrak{L}_{st}$.

**Definition 1.30** (*Kripke-Platek Set Theory* (KP)) The axioms of KP consist of the axioms of BS together with the following $\Delta_0$-collection schema in $\mathfrak{L}_{st}$ [94]:

$$\forall \mathbf{a}(\forall x \exists y \varphi(x, y, \mathbf{a}) \to \forall u \exists v(\forall x \in u)(\exists y \in v)\varphi(x, y, \mathbf{a}))$$

for any $\Delta_0$-formula $\varphi(x, y, \mathbf{a})$.

**Definition 1.31**  A transitive set $M$ is said to be *admissible* if $M \models$ KP. The ordinal $\alpha$ is admissible if $L_\alpha \models$ KP. For any set $X$, $\alpha$ is $X$-admissible if $L_\alpha[X] \models$ KP; $\omega_1^X$ is the least $X$-admissible ordinal and $L_{\omega_1^X}[X]$ is the least admissible set containing $\omega$ and $X$ as elements. A function is $\Sigma_n(L_\alpha)$ if it is $\Sigma_n$ definable over $L_\alpha$ with parameters from $L_\alpha$.

In this book, I make the convention that for admissible ordinal $\alpha$, I always assume that $\alpha > \omega$. The main reference about admissibility is [93].

**Fact 1.14**  (Two basic facts, [92, 94])

*(1)  The ordinal $\alpha$ is $X$-admissible if and only if there is no $\Sigma_1(L_\alpha[X])$ function $f$ which maps some $\beta < \alpha$ cofinally into $\alpha$.[12]*
*(2)  If $A$ is an admissible set and $R \in A$ is a well-ordering, then there exists $\alpha \in A \cap \mathsf{Ord}$, and a function $f \in A$ such that $f$ maps $R$ isomorphically onto $\in \restriction \alpha$.*

For $n_0, \ldots, n_{k-1} \in \omega$, we use $\langle n_0, \ldots, n_{k-1} \rangle$ to denote the natural number encoding $(n_0, \ldots, n_{k-1})$ via a recursive bijection (which we fix throughout) between $\omega^k$ and $\omega$. We regarded reals as codes for relations. Any $x \in \omega^\omega$ encode a binary relation $E_x$ on $\omega$ given by $(m, n) \in E_x \Leftrightarrow x(\langle m, n \rangle) = 0$.

**Definition 1.32**  Let $\mathsf{WF} = \{x \in \omega^\omega \mid E_x$ is well-founded$\}$ and $\mathsf{WO} = \{x \in \omega^\omega \mid E_x$ is well-ordered$\}$. For $x \in \mathsf{WF}$, $rk(x)$ denotes the rank of the well-founded relation $E_x$. For $\alpha < \omega_1$, define $\mathsf{WO}_{<\alpha} = \{x \in \mathsf{WO} \mid rk(x) < \alpha\}$. Define $\omega_1^{\mathsf{CK}} = \sup\{rk(x) : x \in \mathsf{WF}$ and the graph of $x$ is recursive$\}$. $\omega_1^{\mathsf{CK}}$ is the least non-recursive ordinal (similarly, we can define $\omega_1^x$ for real $x$).

**Fact 1.15**  ([92, 93]) *Given a real $x$, $\omega_1^x$ is the least $x$-admissible ordinal, the least admissible ordinal which is not recursive in $x$, the least ordinal which is not the order type of a well-ordering on $\omega$ which is recursive in $x$ and the least ordinal which is not the order type of a $\Delta_1^1(x)$ well-ordering on $\omega$.*

---

[12]Especially, $\alpha$ is admissible if and only if there is no $\Sigma_1(L_\alpha)$ map $f$ which maps some $\beta < \alpha$ cofinally into $\alpha$.

We also use the following notations.

**Definition 1.33** (*Ordinals et cetera*) The set cf($\gamma$) denotes the cofinality of $\gamma$ and $\gamma^+$ denotes the least cardinal greater than $\gamma$. Ord denotes the class of ordinals, $V$ the universe of sets, $V_\alpha$ the set of sets of rank less than $\alpha$. $\overline{X}$ denotes the complement of $X$ and $A \setminus B$ denotes set subtraction. For $X \subseteq$ Ord, $o.t.(X)$ denotes the order type of $X$. For a function $f$, $dom(f)$ denotes its domain, $ran(f)$ its range, $f``X = \{f(y) \mid y \in X\}$, $f \restriction X = f \cap (X \times V)$ (the restriction of $f$ to $X$) and $f^{-1}(X) = \{y \in dom(f) \mid f(y) \in X\}$. If $X, Y$ are sets, then $Y^X$ is the set of all functions from $X$ into $Y$. If $M$ is a transitive set, Ord($M$) denotes Ord $\cap M$. For a set $X$ and cardinal $\kappa$, $[X]^\kappa = \{Y \subseteq X \mid |Y| = \kappa\}$ and $[X]^{<\kappa} = \{Y \subseteq X \mid |Y| < \kappa\}$ (we often write $X^{<\omega}$ for $[X]^{<\omega}$).

**Definition 1.34** (*Limits et cetera*) If $X$ is a set of ordinals and $\alpha > 0$ is a limit ordinal then $\alpha$ is a *limit point* of $X$ if $sup(X \cap \alpha) = \alpha$. Let $\kappa$ be a regular uncountable cardinal. $C \subseteq \kappa$ is *unbounded* in $\kappa$ if $sup(C) = \kappa$; $C$ is closed if it contains all its limit points less than $\kappa$. A set $S \subseteq \kappa$ is *stationary* if $S \cap C \neq \emptyset$ for every closed unbounded subset $C$ of $\kappa$.

In Chaps. 4, 5 and 6, we need the following general notion of club and stationary sets on $[A]^\omega$ for an uncountable set $A$.

**Definition 1.35** ([95]) Let $A$ be an uncountable set and $C \subseteq [A]^\omega$.

  (i) $C$ is unbounded if for any $x \in [A]^\omega$, there is $y \in C$ such that $x \subseteq y$.
  (ii) $C$ is closed if $\bigcup_{n\in\omega} x_n \in C$ for every chain $x_0 \subseteq \cdots \subseteq x_n \subseteq \cdots$ in $C$.
  (iii) $C$ is a club if $C$ is closed and unbounded.
  (iv) $S \subseteq [A]^\omega$ is stationary if $S \cap C \neq \emptyset$ for any club $C$ on $[A]^\omega$.
  (v) For $F : A^{<\omega} \to A$, $x \in [A]^\omega$ is closed under $F$ if $F``(x^{<\omega}) \subseteq x$. Let $C_F = \{x \in [A]^\omega \mid x$ is closed under $F\}$.

Note that if $|A| = \omega_1$, then the concept of club and stationary coincides essentially with the usual concept of club and stationary. We can characterize club and stationary sets in terms of functions $F : A^{<\omega} \to A$.

**Fact 1.16** ([95])

*(1) If $F : A^{<\omega} \to A$, then $C_F$ is a club on $[A]^\omega$.*
*(2) For every club $C$ on $[A]^\omega$, there exists a function $F : A^{<\omega} \to A$ such that $C_F \subseteq C$.*
*(3) $S \subseteq [A]^\omega$ is stationary iff for every function $F : A^{<\omega} \to A$, there is $x \in S$ such that $x$ is closed under $F$.*
*(4) If $A \subseteq B$ and $C$ is a club in $[B]^\omega$, then $C \restriction A = \{x \cap A \mid x \in C\}$ contains a club on $[A]^\omega$.*

**Definition 1.36** A function $f$ defined on a set of ordinals $A$ is *regressive* on $A$ if $f(\alpha) < \alpha$ for all $\alpha \in A$ with $\alpha > 0$. If $X$ is a set of ordinals, a function $f$ defined on $[X]^n([X]^{<\omega})$ is regressive if $f(x) < min(x)$ for all $x \in [X]^n$(resp. all $x \in [X]^{<\omega}$) with $min(x) > 0$.

**Proposition 1.1** (Fodor's theorem, [86]) *If $S \subseteq \kappa$ is stationary, $\kappa$ is uncountable regular cardinal and $f : S \to \kappa$ is regressive, then there is a stationary $\overline{S} \subseteq S$ such that $f$ is constant on $\overline{S}$.*

In this book, the most frequently used notions of large cardinals are $0^\sharp$ and remarkable cardinals. In Sect. 2.1.2, I review some definitions and facts about $0^\sharp$. In Sect. 2.1.3, I review some definitions and facts about remarkable cardinals. For other large cardinal notions in this book, I refer to Appendix C. My basic references for large cardinals are [86, 87].

Finally, I review some basic definitions and facts about forcing that will be used in this book. Our notations about forcing are standard (cf. [86, 95]). For the general theory of forcing, I refer to [86, 91].

**Definition 1.37** A partial order (p.o.) is a partially ordered set $\langle \mathbb{P}, \leq \rangle$ that has a maximum element denoted by $\mathbf{1}$, and for $p, q \in \mathbb{P}$, $p \leq q \Leftrightarrow p$ extends or refines $q$. For a p.o. $\mathbb{P} \in M$, $G$ is $\mathbb{P}$-generic over $M$ if $G \subseteq \mathbb{P}$ is a filter and for any $D \subseteq \mathbb{P}$, if $D$ is dense in $\mathbb{P}$ and $D \in M$, then $G \cap D \neq \emptyset$.

**Definition 1.38** ([86, 91])

(i) A forcing notion $\mathbb{P}$ is $\kappa$-closed if whenever $\lambda < \kappa$ and $\{p_\alpha : \alpha < \lambda\} \subseteq \mathbb{P}$ with $p_\beta \leq p_\alpha$ for $\alpha < \beta < \lambda$, there exists $p \in \mathbb{P}$ such that $p \leq p_\alpha$ for any $\alpha < \lambda$.

(ii) A forcing notion $\mathbb{P}$ satisfies the $\kappa$-chain condition ($\kappa$-c.c.) if every antichain in $\mathbb{P}$ has cardinality less than $\kappa$. $\mathbb{P}$ has countable chain condition(c.c.c.) if it is $\omega_1$-c.c.

(iii) A forcing notion $\mathbb{P}$ is $\kappa$-distributive if the intersection of $\kappa$ open dense sets is open dense. $\mathbb{P}$ is $< \kappa$-distributive if it is $\lambda$-distributive for all $\lambda < \kappa$.

**Fact 1.17** ([86, 91, 95])

*(1) If $(\mathbb{P}, <)$ is $\kappa$-closed, then it is $< \kappa$-distributive.*

*(2) If $\lambda$ is a cardinal and $(\mathbb{P}, <)$ is $\lambda$-closed, then $\mathbb{P}$ preserves cofinality $\leq \lambda$ and hence preserves cardinals $\leq \lambda$.*

*(3) If $\lambda$ is a cardinal and $(\mathbb{P}, <)$ is $\lambda$-c.c, then $\mathbb{P}$ preserves cofinality $\geq \lambda$. If moreover $\lambda$ is a regular cardinal, then $\mathbb{P}$ preserves cardinals $\geq \lambda$.*

*(4) $(\mathbb{P}, <)$ is $\kappa$-distributive iff every function $f : \kappa \to V$ in the generic extension is in the ground model.*

*(5) If $(\mathbb{P}, <)$ is $< \kappa$-distributive, then all cardinals $\leq \kappa$ in $V$ remains cardinals in $V[G]$.*

**Definition 1.39** (*Shelah*, [86, 95]) Suppose $\mathbb{P}$ is a forcing notion, $\kappa > 2^{|\mathbb{P}|}$ is an uncountable cardinal and $M \prec H_\kappa$ such that $|M| = \omega$ and $\mathbb{P} \in M$. We say that a condition $p \in \mathbb{P}$ is $(M, \mathbb{P})$-*generic* if for every dense (antichain, predense) $D \subseteq \mathbb{P}$ with $D \in M$, $D \cap M$ is predense below $p$ (i.e. for all $q \leq p$, there exists $d \in D \cap M$ such that $q$ is compatible with $d$).

**Definition 1.40** (*Shelah*, [86, 95]) A poset $\mathbb{P}$ is *proper* if for every regular uncountable cardinal $\kappa > 2^{|\mathbb{P}|}$, for any $M \prec H_\kappa$ such that $|M| = \omega$ and $\mathbb{P} \in M$, every $p \in \mathbb{P} \cap M$ has an extension $q \leq p$ such that $q$ is an $(M, \mathbb{P})$-generic condition.

**Fact 1.18** (Baumgartner, Jech, Shelah; [86, 95]) *Given a poset $\mathbb{P}$, the following are equivalent to $\mathbb{P}$ being proper:*

*(1) For some regular uncountable cardinal $\kappa > 2^{|\mathbb{P}|}$, for any $M \prec H_\kappa$ such that $|M| = \omega$ and $\mathbb{P} \in M$, every $p \in \mathbb{P} \cap M$ has an extension $q \leq p$ such that $q$ is an $(M, \mathbb{P})$-generic condition;*

*(2) For every uncountable cardinal $\kappa$, $\mathbb{P}$ preserves stationary subsets of $[\kappa]^\omega$;*

*(3) For some (any) regular $\kappa > 2^{|\mathbb{P}|}$, $\{M \prec H_\kappa \mid |M| = \omega, \mathbb{P} \in M$ and $\forall p \in \mathbb{P} \cap M \, \exists q \leq p(q$ is $(M, \mathbb{P})$-generic$)\}$ contains a club subset of $[H_\kappa]^\omega$.*

**Definition 1.41** ([86, 91]) Let $\kappa$ be a regular cardinal and $\lambda > \kappa$ be a cardinal.

(i) $Col(\kappa, \lambda) = \{p : p$ is a function, $dom(p) \subseteq \kappa, |dom(p)| < \kappa$ and $ran(p) \subseteq \lambda\}$.

(ii) $Col(\kappa, < \lambda) = \{p : p$ is a function, $|dom(p)| < \kappa, p \subseteq \lambda \times \kappa$ and $p(\alpha, \beta) < \alpha$ for any $(\alpha, \beta) \in dom(p)\}$.

(iii) $Fn(I, J, \kappa) = \{p : p$ is a function, $dom(p) \subseteq I, ran(p) \subseteq J$ and $|dom(p)| < \kappa\}$.[13]

(iv) In all these forcing notions, $p \leq q$ if $p \supseteq q$.

**Fact 1.19** ([86, 91])

*(i) $Col(\kappa, \lambda)$ collapses $\lambda$ to $\kappa$. $Col(\kappa, < \lambda)$ collapses any $\kappa < \alpha < \lambda$ to $\kappa$ and collapses $\lambda$ to $\kappa^+$.*

*(ii) $Col(\kappa, \lambda)$ and $Col(\kappa, < \lambda)$ are $\kappa$-closed.*

*(iii) $Fn(I, J, \kappa)$ is $(|J|^{<\kappa})^+$-c.c. If $\kappa$ is regular, then $Fn(I, J, \kappa)$ is $\kappa$-closed. Especially, if $\kappa^{<\gamma} = \kappa$, then $Fn(\gamma, \kappa, \gamma)$ is $\kappa^+$-c.c.*

*(iv) If $\lambda$ is regular and for any $\alpha < \lambda, \alpha^{<\kappa} < \lambda$, then $Col(\kappa, < \lambda)$ is $\lambda$-c.c.*

**Theorem 1.20** (Levy, [86]) *Suppose $\kappa$ is a regular cardinal, $\lambda > \kappa$ is an inaccessible cardinal and $G$ is $Col(\kappa, < \lambda)$-generic over $V$.*

*(a) Every $\alpha$ such that $\kappa \leq \alpha < \lambda$ has cardinality $\kappa$ in $V[G]$.*

*(b) Every cardinal $\leq \kappa$ and every cardinal $\geq \lambda$ remains a cardinal in $V[G]$.*

*Hence, $V[G] \models \lambda = \kappa^+$.*

*Proof* By Facts 1.19 and 1.17, $Col(\kappa, < \lambda)$ is $\lambda$-c.c. and $\kappa$-closed.

The following theorem introduces the forcing technique called reshaping. In Sects. 2.3, 4.3 and 4.5, I use variants of the reshaping technique.

**Theorem 1.21** (Beller, Jensen and Welch, [96]) *Let $\gamma$ be an infinite cardinal, and let $B \subseteq \gamma^+$ such that $V = L[B]$. Then we can generically add a $D \subseteq \gamma$ such that*

*(1) $B \in L[D]$,*

*(2) $\forall \delta < \gamma^+ (L[D \cap \delta] \models |\delta| \leq \gamma)$, and*

*(3) $L[D]$ preserves cardinals and cofinalities.*

---

[13]Note that $Col(\gamma, \kappa) = Fn(\gamma, \kappa, \gamma)$.

## 1.2  Introduction

In this section, I introduce my research problems and outline the structure of this book.

As an easy corollary of G2, if $Z_2$ is consistent, then there is a true sentence about analysis that is not provable in $Z_2$. Many classic mathematical theorems about analysis expressible in SOA are provable in $Z_2$. The following is the motivation question for the below:

*Question 1.1* Are all theorems in classic mathematics expressible in SOA provable in $Z_2$? Or is $Z_2$ complete for classical mathematics expressible in SOA?

As discussed in Sect. 1.1.3, people have found many examples of concrete mathematical theorems which are not provable in SOA. In this book, I give a new counterexample to Question 1.1 which is isolated from a famous theorem in Set Theory: the *Martin-Harrington Theorem*.

Over the last four decades, much work has been done on the relationship between large cardinal and the determinacy hypotheses, especially the large cardinal-determinacy correspondence. The first result in this line was proved by Martin and Harrington, as follows.

**Theorem 1.22** (Martin, [86]) *In* ZF, *the existence of* $0^\sharp$ *implies* $Det(\Sigma_1^1)$.

**Theorem 1.23** (Harrington, [97]) *In* ZF, $Det(\Sigma_1^1)$ *implies that* $0^\sharp$ *exists.*

The Martin-Harrington Theorem is a milestone for the correspondence between large cardinal and the determinacy hypotheses. This remarkable equivalence is an unexpected confluence that bolstered both large cardinal and the determinacy theory, and motivated further research on the relationship between large cardinal and determinacy hypotheses.

Throughout this book, I make the following convention: Martin's Theorem is the statement "$0^\sharp$ exists implies $Det(\Sigma_1^1)$"; the **Boldface** Martin Theorem is the statement "for any real $x$, $x^\sharp$ exists implies $Det(\boldsymbol{\Sigma}_1^1)$"; the Harrington's Theorem is the statement "$Det(\Sigma_1^1)$ implies that $0^\sharp$ exists"; the **Boldface** Harrington Theorem is the statement "$Det(\boldsymbol{\Sigma}_1^1)$ implies that for any real $x$, $x^\sharp$ exists"; the Martin-Harrington Theorem is the statement "$Det(\Sigma_1^1)$ if and only if $0^\sharp$ exists"; the **Boldface** Martin-Harrington Theorem is the statement "$Det(\boldsymbol{\Sigma}_1^1)$ iff for any real $x$, $x^\sharp$ exists".

In Chap. 3, I show that the Boldface Martin-Harrington Theorem is provable in $Z_2$. We know that the Martin-Harrington Theorem is provable in ZF and expressible in SOA. In fact, both $Det(\Sigma_1^1)$ and "$0^\sharp$ exists" are $\Sigma_3^1$. A natural question is then:

*Question 1.2* Is the Martin-Harrington Theorem provable in $Z_2$?

I show in Sect. 3.1 that Martin's Theorem is provable in $Z_2$. After that, I focus on the analysis of Harrington's Theorem in higher-order arithmetic. Indeed, Harrington isolated an important principle in the proof of Theorem 1.23: *Harrington's Principle*, defined as follows.

**Definition 1.42** We let *Harrington's Principle*, HP for short, denote the following statement: $(\exists x \in \omega^\omega)(\forall \alpha)$ (if $\alpha$ is a countable $x$-admissible ordinal, then $\alpha$ is an $L$-cardinal).

**Theorem 1.24** (Silver, [97]) *In* ZF, HP *implies that* $0^\sharp$ *exists.*

**Theorem 1.25** (Silver, Solovay, [86]) *In* ZF, $0^\sharp$ *exists implies* HP*: Assuming* $0^\sharp$ *exists and $I$ is the class of Silver indiscernibles, if $\alpha$ is $0^\sharp$-admissible, then $\alpha \in I$ and hence $\alpha$ is an $L$-cardinal.*

In a summary, in ZF we have $Det(\Sigma_1^1) \Leftrightarrow$ HP $\Leftrightarrow 0^\sharp$ exists. There is no direct proof of Harrington's Theorem as far as I know. All known proofs of Harrington's Theorem use the intermediate statement HP and are done in two steps:

First Step:     $Det(\Sigma_1^1) \Rightarrow$ HP;
Second Step:    HP $\Rightarrow 0^\sharp$ exists.

Now, a *transfer theorem* is a result of the form $\phi \Rightarrow \psi$ such that there is no direct proof known for it: all known proofs of it use an intermediate statement and are done in two steps: first show that $\phi \Rightarrow \varphi$ and then show that $\varphi \Rightarrow \psi$ for some intermediate statement $\varphi$. Harrington's Theorem 1.23 is a typical example of transfer theorems in the study of the correspondence between the determinacy hypotheses and large cardinals. Before we move on, in the following I give some other examples of transfer theorems in Set Theory.

**Theorem 1.26** (Harrington, Martin; 1970s) $Det(\Pi_1^1)$ *implies* $Det(< \omega^2\text{-}\Pi_1^1)$.

We do not know a direct proof of Theorem 1.26: all known proofs are done in the following two steps:

**Theorem 1.27**

*(1) (Harrington)* $Det(\Pi_1^1)$ *implies that* $0^\sharp$ *exists.*
*(2) (Martin)* $0^\sharp$ *exists implies* $Det(< \omega^2\text{-}\Pi_1^1)$.

For a pointclass $\Gamma$, a set of reals is in $\partial^n(\Gamma)$ if it can be defined from a set in $\Gamma$ using $n$ successive applications of $\partial$ where $(\partial y)A(x, y) \Leftrightarrow I$ wins $G(\{y|(x, y) \in A\})$ (see [98]). The following theorem due to Woodin and Neeman is a generalization of Theorem 1.26.

**Theorem 1.28** (Neeman, Woodin; 1990s) $Det(\Pi_{n+1}^1)$ *implies* $Det(\partial^n(< \omega^2\text{-}\Pi_1^1))$ *where $n \in \omega$.*[14]

All known proofs of Theorem 1.28 are done in the following two steps. For the definition of *Woodin cardinals*, I refer to Definition C.15 below.

---

[14]Note that there are more sets in $\partial^n(< \omega^2\text{-}\Pi_1^1)$ than there are in $\Pi_{n+1}^1$.

**Theorem 1.29**

(1) *(Woodin) $Det(\Pi^1_{n+1})$ implies that there are certain iterable inner models with n Woodin cardinals.*

(2) *(Neeman) If $Det(\Pi^1_1)$ holds, then the conclusion of the above Woodin's theorem yields $Det(\partial^n(< \omega^2\text{-}\Pi^1_1))$.*

The proof of Theorem 1.28 uses the following remarkable result. For the definition of $M^\sharp_n(x)$, I refer to Definition C.16.

**Theorem 1.30** (Martin, Steel [99], Woodin [100]) *The following are equivalent:*

(1) *for all $n \in \omega$, $Det(\Pi^1_n)$ holds;*

(2) *for all $n \in \omega$, for all $x \in \omega^\omega$, $M^\sharp_n(x)$ exists.*

As Theorem 1.30 shows, the work of Martin, Steel and Woodin has precisely characterized the large cardinal assumptions needed to prove each level of determinacy in the projective hierarchy. Theorem 1.30 provides a route to prove transfer theorems of the form $\phi \Rightarrow$ PD: by Theorem 1.30, the intermediate statement will be "for all $n \in \omega$, for all $x \in \omega^\omega$, $M^\sharp_n(x)$ exists".[15]

Now, let us return to the analysis of Theorem 1.23. Recall that all known proofs of Harrington's Theorem "$Det(\Sigma^1_1)$ implies that $0^\sharp$ exists" in ZF are done in two steps. We observe that the first step "$Det(\Sigma^1_1)$ implies HP" is provable in $Z_2$ based on known proofs of Harrington's Theorem. In the following, I give a brief survey of different proofs of "$Z_2 + Det(\Sigma^1_1)$ implies HP" in the literature.

There are several published proofs of "$Z_2 + Det(\Sigma^1_1)$ implies HP" in the literature (see Harrington [97], Friedman [92], Sami [101] and Schindler [102]). In particular, Schindler gives a detailed proof of Harrington's theorem in [102, pp. 296–300] using some basic fine structure of $L$ and Steel's forcing. Proofs in [92, 97, 101] in fact show that $Z_2 + Det$(Turing-$\Sigma^1_1$) implies HP. Proofs in [92, 97, 102] use Steel forcing. Woodin also gave a proof of "$Z_2 + Det(\Sigma^1_1)$ implies HP" using only basic Cohen forcing $Col(\omega, \alpha)$ without the use of Steel forcing.[16] Sami's proof in [101] is the only forcing-free proof I know which uses only effective descriptive set theory. For Sami's forcing-free proof of "$Z_2 + Det$(Turing-$\Sigma^1_1$) implies HP", I refer to Appendix A.

Now, since "$Det(\Sigma^1_1)$ implies HP" is provable in $Z_2$, a natural question is:

*Question 1.3* Does $Z_2 + $ HP imply that $0^\sharp$ exists?

If the answer is positive, then Harrington's Theorem is provable in $Z_2$. If the answer is negative, and Harrington's Theorem is provable in $Z_2$, then there must be a new proof of Harrington's Theorem without the use of HP to derive the existence of $0^\sharp$. In joint work with Ralf Schindler, we prove in Sect. 2.2 that $Z_2 + $ HP is equiconsistent with ZFC. As a corollary, $Z_2 + $ HP does not imply that $0^\sharp$ exists. Then the next natural question is:

---

[15]That is, from $\phi$ we will have to show that the mouse operator $x \to M^\sharp_n(x)$ is total for all $n \in \omega$. Woodin's *core model induction* is one key method to achieve this: by induction on $n$. For more details on transfer theorems and Woodin's core model induction, I refer to [98].

[16]Woodin's proof is unpublished. For a reconstruction of Woodin's proof, I refer to [103].

*Question 1.4*  Does $Z_3 + HP$ imply that $0^\sharp$ exists?

If the answer is positive, then $Z_3$ is the minimal system of higher-order arithmetic to prove "HP implies that $0^\sharp$ exists" and Harrington's Theorem is provable in $Z_3$. We prove in Sect. 2.3 that $Z_3 + HP$ is equiconsistent with ZFC+ there exists a remarkable cardinal. As a corollary, $Z_3 + HP$ does not imply that $0^\sharp$ exists.

Next, one can ask whether $Z_4 + HP$ implies that $0^\sharp$ exists? In Sect. 2.4, I prove that $Z_4 + HP$ indeed implies that $0^\sharp$ exists. Hence, $Z_4$ is the minimal system in higher-order arithmetic to prove "HP implies that $0^\sharp$ exists". This is the first example I know that $Z_4$ is the minimal system in higher-order arithmetic to prove a concrete mathematical theorem expressible in SOA.

Finally, as to the structure of the book, it consists of Chaps. 1–6 and Appendix A-C.

- In Chap. 1, I give an overview of Incompleteness, Reverse Mathematics and Incompleteness for higher-order arithmetic, review some notions and facts in Set Theory used in this book, and introduce my research problems and outline the structure of this book.
- In Chap. 2, we show that $Z_2 + HP$ is equiconsistent with ZFC and that $Z_3 + HP$ is equiconsistent with "ZFC+ there exists a remarkable cardinal". As a corollary, "HP implies that $0^\sharp$ exists" is neither provable in $Z_2$ nor provable in $Z_3$. Finally, I show that $Z_4 + HP$ implies that $0^\sharp$ exists. Thus, $Z_4$ is the minimal system to show that Harrington's Principle implies that $0^\sharp$ exists in higher-order arithmetic.
- In Chap. 3, I prove the Boldface Martin-Harrington Theorem in $Z_2$.
- In Chap. 4, we examine the large cardinal strength of strengthenings of Harrington's Principle over $Z_2$ and $Z_3$.
- In Chap. 5, I force a model of "$Z_3$+ Harrington's Principle" via set forcing without the use of the reshaping technique, but assuming there exists a remarkable cardinal with a weakly inaccessible cardinal above it.
- In Chap. 6, I develop the full theory of the strong reflecting property for $L$-cardinal and characterize the strong reflecting property of $\omega_n$ for $n \in \omega$.
- In Appendix $A$, I reconstruct Sami's forcing-free proof of "$Det(Turing\text{-}\Sigma_1^1)$ implies HP" in $Z_2$ in a general setting.
- In Appendix $B$, I show that $Z_3 + Det(< \omega^2\text{-}\Pi_1^1)$ implies that $0^\sharp$ exists.
- In Appendix $C$, I review some definitions of large cardinals used in this book that are not covered in Sects. 2.1.2 and 2.1.3.

This book contributes to the research program on concrete incompleteness for higher-order arithmetic and gives a specific example of concrete mathematical theorems which is expressible in second-order arithmetic but the minimal system in higher-order arithmetic to prove it is $Z_4$. The question "whether Harrington's Theorem is provable in $Z_2$" is still open as far as I know. W.Hugh Woodin conjectured that Harrington's Theorem is provable in $Z_2$.

# References

1. Feferman, S.: The impact of the incompleteness theorems on mathematics. Notices AMS. **53**(4), 434–439 (2006)
2. Enderton, H.B.: A Mathematical Introduction to Logic, 2nd edn. Academic Press, Boston, MA (2001)
3. Murawski, R.: Recursive Functions and Metamathematics: Problems of Completeness and Decidability. Gödel's Theorems. Springer, Netherlands (1999)
4. Lindström, P.: Aspects of Incompleteness. Lecture Notes in Logic v. **10** (1997)
5. Smith, P.: An Introduction to Gödel's Theorems. Cambridge University Press (2007)
6. Smullyan, M.R.: Gödel's Incompleteness Theorems. Oxford Logic Guides 19. Oxford University Press (1992)
7. Smullyan, M.R.: Diagonalization and Self-Reference. Oxford Logic Guides 27. Clarendon Press (1994)
8. Boolos, G.: The Logic of Provability. Cambridge University Press (1993)
9. Beklemishev, L.D.: Gödel incompleteness theorems and the limits of their applicability I. Russ. Math. Surv. (2010)
10. Kotlarski, H.: The incompleteness theorems after 70 years. Ann. Pure Appl. Logic. **126**, 125–138 (2004)
11. Smoryński, C.: The incompleteness theorems. In: Barwise, J. (ed.) Handbook of Mathematical Logic, pp. 821–865. North-Holland, Amsterdam (1977)
12. Visser, A.: The second incompleteness theorem: reflections and ruminations. Chapter in Gödel's Disjunction: The scope and limits of mathematical knowledge, edited by Leon Horsten and Philip Welch, Oxford University Press (2016)
13. Hájek, P., Pudlák, P.: Metamathematics of First-Order Arithmetic. Springer, Berlin, Heidelberg, New York (1993)
14. Visser, A.: Can we make the second incompleteness theorem coordinate free? J. Logic Comput. **21**(4), 543–560 (2011)
15. Visser, A.: Why the theory R is special. In: Tennant, N. (Eds.), Foundational Adventures-Essays in Honour of Harvey M. Friedman, pp. 7–24. (17 p.). College Publication (2014)
16. Gödel, K.: Über formal unentscheidbare Sätze der Principia Mathematica und verwandter Systeme I. Monatsh. Math. Phys. **38**(1), 173–198 (1931)
17. Zach, R.: Hilbert's Program Then and Now. Philosophy of Logic, Handbook of the Philosophy of Science, pp. 411–447 (2007)
18. Tarski, A., Mostowski, A., Robinson, M.R.: Undecidable Theories. North-Holland (1953)
19. Kikuchi, M., Kurahashi, T.: Generalizations of Gödel's incompleteness theorems for $\Sigma_n$-definable theories of arithmetic. Rew. Symb. Logic **10**(4), 603–616 (2017)
20. Salehi, S., Seraji, P.: Gödel-Rosser's Incompleteness Theorem, generalized and optimized for definable theories. J. Logic Comput. **27**(5), 1391–1397 (2017)
21. Jones, J.P., Shepherdson, J.C.: Variants of Robinson's essentially undecidable theory $R$. Arch. Math. Logic **23**, 65–77 (1983)
22. Vaught, L.R.: On a theorem of Cobham concerning undecidable theories. In: Nagel, E., Suppes, P., Tarksi, A. (Eds.): Logic, Methodology, and Philosophy of Science, p. 18. Proceedings of the 1960 International Congress. Stanford, CA: Stanford University Press (1962)
23. Cheng, Y.: Finding the limit of incompleteness I. Preprint. See http://yongcheng.whu.edu.cn/
24. Kikuchi, M., Kurahashi, T.: Universal Rosser predicates. J. Symb. Log. **82**(1), 292–302 (2017)
25. Feferman, S.: Arithmetization of mathematics in a general setting. Fundamenta Mathematicae. **49**, 35–92 (1960)
26. Detlefsen, M.: On a theorem of Feferman. Philos. Stud.: Int. J. Philos. Anal. Tradit. **38**(2), 129–140 (1980)
27. Franks, C.: The Autonomy of Mathematical Knowledge: Hilbert's Program Revisited. Cambridge University Press (2009)
28. Detlefsen, M.: On interpreting Gödel's second theorem. J. Philos. Logic **8**, 297–313 (1979)

29. Artemov, S.: The Provability of Consistency. Available at arXiv:1902.07404v1
30. Friedman, H.: Some systems of second-order arithmetic and their use. In: Proceedings of the International Congress of Mathematicians, Vol. 1, 1975, pp. 235–242 (1974)
31. Friedman, H.: Systems of second-order arithmetic with restricted induction, I and II (Abstracts). J. Symb. Log. **41**, 557–559 (1976)
32. Simpson, G.S.: Subsystems of second-order arithmetic, Perspectives in Logic (2nd ed.), Cambridge University Press (2009)
33. Stillwell, J.: Reverse Mathematics, Proofs From the Inside Out. Princeton University Press (2018)
34. Heijenoort Van, J.: From Frege to Gödel: A Source Book in Mathematical Logic 1879–1931. Harvard University Press (1967)
35. Normann, D., Sanders, S.: On the mathematical and foundational significance of the uncountable. arXiv:https://arxiv.org/abs/1711.08939 (2017)
36. Reverse Mathematics Zoo. See https://rmzoo.math.uconn.edu/
37. Simpson, S.G.: The Gödel Hierarchy and Reverse Mathematics in Kurt Gödel: Essays for his Centennial (Lecture Notes in Logic 33) Feferman, S., Parsons, C., Simpson, S.G. (Eds.). Cambridge University Press (2010)
38. Aberth, O.: Computable Analysis. McGraw-Hill (1980)
39. Pour-El, M.B., Richards, J.I.: Computability in Analysis and Physics. Perspectives in Mathematical Logic. Springer, XI + 206 p. (1988)
40. Hilbert, D.: Über das Unendliche. Mathematische Annalen **95**, 161–190 (1926)
41. Weyl, H.: The Continuum: A Critical Examination of the Foundations of Analysis. Thomas Jefferson University Press (1987)
42. Kreisel, G.: The axiom of choice and the class of hyperarithmetic functions. Indagationes Mathematicae, **24**, 307–319 (1962)
43. Feferman, S.: Systems of predicative analysis I. J. Symb. Log. **29**, 1–30 (1964)
44. Feferman, S.: Systems of predicative analysis II. J. Symb. Log. **33**, 193–220 (1968)
45. Feferman, S.: Weyl vindicated: Das Kontinuum seventy years later. In: The Light of Logic, pp. 249–283 (1998)
46. Feferman, S.: Predicatively reducible systems of set theory. In: Number XIII in Proceedings of Symposia in Pure Mathematics, pp. 11–32 (1974)
47. Simpson, S.G.: Predicativity: the outer limits. In: Reflections on the Foundations of Mathematics: Essays in Honor of Solomon Feferman, pp. 134–140 (2002)
48. Friedman, H., McAloon, K., Simpson, S.G.: A finite combinatorial principle which is equivalent to the 1-consistency of predicative analysis. In: Patras Logic Symposion Studies in Logic and the Foundations of Mathematics, pp. 197–230 (1982)
49. Buchholz, W., Feferman, S., Pohlers, W., Sieg, W.: Iterated Inductive Definitions and Subsystems of Analysis: Recent Proof-Theoretical Studies. Number 897 in Lecture Notes in Mathematics. Springer (1981)
50. Kohlenbach, U.: Higher-order reverse mathematics, pp. 281–295, in Lecture Note In Logic 21: Reverse Mathematics 2001. Association of Symbolic Logic (2005)
51. Kohlenbach, U.: Foundational and mathematical uses of higher types. Reflect. Found. Math. Lect. Notes Log. **15**, ASL, 92–116 (2002)
52. Normann, D., Sanders, S.: Uniformity in Mathematics. (2018)
53. Gitman, V., Hamkins, J.D., Johnstone, T.A.: What is the theory **ZFC** without Powerset? Math Logic Quart. **62**(4–5), 391–406 (2016)
54. Montalbán, A., Shore, R.A.: The limits of determinacy in second-order Arithmetic. Proc. London Math Soc. **104**, 223–252 (2012)
55. Montalbán, A., Shore, R.A.: The limits of determinacy in second-order arithmetic: consistency and complexity strength. Israel J. Math. **204**, 477–508 (2014)
56. Bovykin, A.: Brief introduction to unprovability. Logic Colloquium 2006, Lecture Notes in Logic 32
57. Paris, J., Harrington, L.: A mathematical incompleteness in Peano arithmetic. In: Barwise, J. (Ed.) Handbook of Mathematical Logic, Studies in Logic and the Foundations of Mathematics, vol. 90, pp. 1133–1142. North-Holland, Amsterdam, New York, Oxford (1977)

58. Graham, R.L., Rothschild, B.L., Spencer, J.H.: Ramsey Theory. John Wiley and Sons (1990)
59. Erdos, P., Rado, R.: Combinatorial theorems on classifications of subsets of a given set. Proc. London Math Soc. III Ser. **2**, 417–439. MR 16:455d (1952)
60. Kanamori, A., McAloon, K.: On Gödel's incompleteness and finite combinatorics. Ann. Pure Appl. Logic. **33**(1), 23–41 (1987)
61. Paris, J., Kirby, L.: Accessible independence results for Peano arithmetic. Bull. London Math. Soc. **14**(4), 285–293 (1982)
62. Beklemishev, L.D.: The Worm principle. Logic Group Preprint Series 219, Utrecht University (2003)
63. Hamano, M., Okada, M.: A relationship among Gentzen's proof-reduction, Kirby-Paris' hydra game, and Buchholz's hydra game. Math. Logic Quart. **43**(1), 103–120 (1997)
64. Kirby, L.: Flipping properties in arithmetic. J. Symb. Log. **47**(2), 416–422 (1982)
65. Mills, G.: A tree analysis of unprovable combinatorial statements. Model Theory of Algebra and Arithmetic, Lecture Notes in Mathematics, vol. 834, pp. 248–311. Springer, Berlin (1980)
66. Pudlak, P.: Another combinatorial principle independent of Peano's axioms. Unpublished (1979)
67. Hájek, P., Paris, J.: Combinatorial principles concerning approximations of functions. Archive Math. Logic. **26**(1–2), 13–28 (1986)
68. Clote, P., McAloon, K.: Two further combinatorial theorems equivalent to the 1-consistency of Peano arithmetic. J. Symb. Log. **48**(4), 1090–1104 (1983)
69. Kanamori, A., McAloon, K.: On Gödel incompleteness and finite combinatorics. Ann. Pure Appl. Logic. **33**, 23–41 (1987)
70. Friedman, H.: Boolean relation theory and incompleteness. Manuscript, to appear
71. Friedman, H.: Finite functions and the neccessary use of large cardinals. Ann. Math. **148**, 803–893 (1998)
72. Friedman, H.: On the necessary use of abstract set theory. Adv. Math. **41**, 209–280 (1981)
73. Simpson, G.S. (Eds.): Nonprovability of certain combinatorial properties of finite trees. Harvey Friedman's Research on the Foundations of Mathematics, Studies in Logic and the Foundations of Mathematics, vol. 117, North-Holland,Amsterdam, pp. 87–117 (1985)
74. Friedman, H., Robertson, N., Seymour, P.: The metamathematics of the graph minor theorem. Contemp. Math. **65**, 229–261 (1987)
75. Krombholz, M.R.A.: Proof theory of graph minors and tree embeddings. Ph.D. Thesis, University of Leeds (2018)
76. Simpson, G.S. (Ed.): Logic and combinatorics. Contemp. Math. **65** AMS, Providence, RI (1987)
77. Simpson, G.S.: Harvey Friedman's Research on the Foundations of Mathematics, Studies in Logic and the Foundations of Mathematics, vol. 117. North-Holland Publishing, Amsterdam (1985)
78. Pacholski, L., Wierzejewski, J.: Model Theory of Algebra and Arithmetic. Lecture Notes in Mathematics, vol. 834. Springer, Berlin (1980)
79. Berline, C., McAloon, K., Ressayre, J.P. (Eds.): Model Theory and Arithmetic. Lecture Notes in Mathematics, vol. 890. Springer, Berlin (1981)
80. Weiermann, A.: An application of graphical enumeration to PA. J. Symb. Log. **68**(1), 5–16 (2003)
81. Weiermann, A.: A classification of rapidly growing Ramsey functions. Proc. Am. Math. Soc. **132**, 553–561 (2004)
82. Weiermann, A.: Analytic combinatorics, proof-theoretic ordinals, and phase transitions for independence results. Ann. Pure Appl. Logic. **136**, 189–218 (2005)
83. Weiermann, A.: Classifying the provably total functions of $PA$. Bull. Symb. Logic **12**(2), 177–190 (2006)
84. Weiermann, A.: Phase transition thresholds for some Friedman-style independence results. Math. Logic Quart. **53**(1), 4–18 (2007)
85. Gordeev, L., Weiermann, A.: Phase transitions of iterated Higman-style well-partial-orderings. Archive Math. Logic. **51**(1–2), 127–161 (2012)

86. Jech, T.J.: Set Theory. Third Millennium Edition, revised and expanded. Springer, Berlin (2003)

87. Kanamori, A.: Higher Infinite: Large Cardinals in Set Theory from Their Beginnings, 2nd edn. Springer Monographs in Mathematics, Springer, Berlin (2003)

88. Friedman, H.: Higher set theory and mathematical practice. Ann. Math. **2**, 325–357 (1971)

89. Martin, D.A.: Borel determinacy. Ann. Math. **102**, 363–371 (1975)

90. Martin, D.A.: $\Sigma_4^0$-determinacy, circulated handwritten notes dated March **20** (1974)

91. Kunen, K.: Set Theory: An Introduction to Independence Proofs. North Holland (1980)

92. Mansfield, R., Weitkamp, G.: Recursive Aspects of Descriptive Set Theory. Oxford University Press, Oxford (1985)

93. Barwise, J.: Admissible Sets and Structures. Perspectives in Math. Logic Vol. 7, Springer (1976)

94. Devlin, K.J.: Constructibility. Springer, Berlin (1984)

95. Jech, T.J.: Multiple Forcing. Cambridge University Press (1986)

96. Beller, A., Jensen, R.B., Welch, P.: Coding the Universe. Cambridge University Press, Cambridge (1982)

97. Harrington, L.A.: Analytic determinacy and $0^{\sharp}$. J. Symb. Log. **43**, 685–693 (1978)

98. Schindler, R., Steel, J.: The core model induction. Manuscipt

99. Martin, D.A., Steel, J.R.: A proof of projective determinacy. J. Am. Math. Soc. **2**, 71–125 (1989)

100. Woodin, W.H.: Supercompact cardinals, sets of reals, and weakly homogeneous trees. In: Proceedings of the National Academy of Sciences of the U.S.A., vol. 85, pp. 6587–6591 (1988)

101. Sami, Ramez, L.: Analytic determinacy and $0^{\sharp}$: A forcing-free proof of Harrington's theorem. Fundamenta Mathematicae **160** (1999)

102. Schindler, R.: Set Theory: Exploring Independence and Truth. Springer (2014)

103. Cheng, Y.: Analysis of Martin-Harrington theorem in higher-order arithmetic. Ph.D. thesis, National University of Singapore (2012)

# Chapter 2
# A Minimal System

**Abstract** In this chapter, we prove the following results.

(1) $Z_2 + HP$ is equiconsistent with ZFC.
(2) $Z_3 + HP$ is equiconsistent with ZFC+ "there exists a remarkable cardinal".
(3) $Z_4 + HP$ implies that $0^\sharp$ exists.

As a corollary, "HP implies that $0^\sharp$ exists" is neither provable in $Z_2$ nor in $Z_3$, i.e. $Z_4$ is the minimal system of higher-order arithmetic for proving that "HP implies that $0^\sharp$ exists".

## 2.1 Preliminaries

In this chapter, I determine the minimal system of higher-order arithmetic for proving that HP implies that $0^\sharp$ exists. The structure of this chapter is as follows. In Sect. 2.1, I introduce almost disjoint forcing, $0^\sharp$, and remarkable cardinals. In Sect. 2.2, we prove that $Z_2 + HP$ is equiconsistent with ZFC. As a corollary, $Z_2 + HP$ does not imply that $0^\sharp$ exists. In Sect. 2.3, we prove that $Z_3 + HP$ is equiconsistent with ZFC + there is a remarkable cardinal. As a corollary, $Z_3 + HP$ does not imply that $0^\sharp$ exists. In Sect. 2.4, I prove that $Z_4 + HP$ implies that $0^\sharp$ exists. Thus, $Z_4$ is the minimal system of higher-order arithmetic for proving that HP implies that $0^\sharp$ exists.

### 2.1.1 Almost Disjoint Forcing

Almost disjoint forcing is used in Sects. 2.2, 2.3, 4.3, 4.5 and Chap. 5. In this section, I introduce almost disjoint forcing and examine the required basic properties. I refer to [1–3] for further details.

**Definition 2.1** ([1, 2]) The class $\mathscr{F} = \{a_\alpha : \alpha < \lambda\}$ is an *almost disjoint family of size $\lambda$ on $\kappa$* if

(1)  for all $\alpha < \lambda$, $a_\alpha \subseteq \kappa$ and $|a_\alpha| = \kappa$;
(2)  for all $\alpha, \beta$ in $\lambda$ with $\alpha \neq \beta$, we have $|a_\alpha \cap a_\beta| < \kappa$.

Suppose $\kappa$ is a regular cardinal, $\lambda > \kappa$ and $\mathscr{F} = \{a_\alpha : \alpha < \lambda\}$ is an almost disjoint family of size $\lambda$ on $\kappa$. Given $A \subseteq \lambda$, using $\mathscr{F}$ we can force to add a $B \subseteq \kappa$ such that $B$ codes $A$ in the following sense:

$$A = \{\alpha < \lambda : |B \cap a_\alpha| < \kappa\}.$$

This observation gives rise to the following further definition.

**Definition 2.2** ([1, 2]) Given $\mathscr{F}$ and $A$ as above, we define the almost disjoint forcing notion $\mathbb{P}_{\mathscr{F},A}$ as follows:

$$\mathbb{P}_{\mathscr{F},A} = [\kappa]^{<\kappa} \times [A]^{<\kappa};$$

$$(p, q) \leq (p', q') \Leftrightarrow (p \supseteq p' \wedge q \supseteq q' \wedge \forall \alpha \in q'(p \cap a_\alpha \subseteq p')).$$

For $\alpha \in A$, define $D_\alpha = \{(p, q) \in \mathbb{P}_{\mathscr{F},A} \mid \alpha \in q\}$. For $\alpha < \kappa$ and $\beta \in \lambda \setminus A$, define $D_{\alpha,\beta} = \{(p, q) \in \mathbb{P}_{\mathscr{F},A} \mid o.t.(p \cap a_\beta) \geq \alpha\}$. For $\alpha \in A$, $D_\alpha$ is dense in $\mathbb{P}_{\mathscr{F},A}$ since for any $(p, q) \in \mathbb{P}_{\mathscr{F},A}$, $(p, q \cup \{\alpha\}) \leq (p, q)$ and $(p, q \cup \{\alpha\}) \in D_\alpha$.

**Lemma 2.1** *For $\alpha < \kappa$ and $\beta \in \lambda \setminus A$, $D_{\alpha,\beta}$ is dense in $\mathbb{P}_{\mathscr{F},A}$.*

*Proof* Given $(p, q) \in \mathbb{P}_{\mathscr{F},A}$, let $S = a_\beta \setminus \bigcup_{\gamma \in q}(a_\beta \cap a_\gamma)$. Since $\beta \notin A$, for $\gamma \in q \subseteq A$, we have $|a_\beta \cap a_\gamma| < \kappa$. Since $|q| < \kappa$, we have $|S| = \kappa$. Let $p' = p \cup D$ with $D \subseteq S$ and $o.t.(D) = \alpha$. So $o.t.(p' \cap a_\beta) \geq \alpha$. By the definition of $S$, we have $p' \cap a_\gamma \subseteq p$ for all $\gamma \in q$. Thus, $(p', q) \leq (p, q)$ and $(p', q) \in D_{\alpha,\beta}$. $\square$

**Theorem 2.1** *Let $G$ be $\mathbb{P}_{\mathscr{F},A}$-generic over $V$. Let $B = \bigcup\{p \mid \exists q\,((p, q) \in G)\}$. Then we have*
$$A = \{\alpha < \lambda : |B \cap a_\alpha| < \kappa\}.$$

*Proof* If $\alpha \in A$, since $D_\alpha$ is dense, there is $(p, q) \in G$ such that $\alpha \in q$. So $B \cap a_\alpha \subseteq p$ and $|B \cap a_\alpha| \leq |p| < \kappa$. If $\beta \in \lambda \setminus A$, then $D_{\alpha,\beta}$ is dense for all $\alpha < \kappa$. Thus, there is $(p, q) \in G$ such that $o.t.(p \cap a_\beta) \geq \alpha$. Since $o.t.(p \cap a_\beta) \geq \alpha$ for any $\alpha < \kappa$, we have $|B \cap a_\alpha| = \kappa$. $\square$

**Proposition 2.1** $\mathbb{P}_{\mathscr{F},A}$ *is $\kappa$-closed.*

*Proof* Let $\alpha < \kappa$ and $\langle (p_\xi, q_\xi) : \xi < \alpha \rangle$ be a descending sequence of conditions. Let $p' = \bigcup_{\xi < \alpha} p_\xi$ and $q' = \bigcup_{\xi < \alpha} q_\xi$. Note that for $\xi < \alpha$ and $\beta \in q_\xi$, we have $p' \cap a_\beta \subseteq p_\xi$. Thus, for any $\xi < \alpha$, we have $(p', q') \leq (p_\xi, q_\xi)$. Also since $\kappa$ is regular, we have $(p', q') \in \mathbb{P}_{\mathscr{F},A}$. $\square$

**Proposition 2.2** *If $\kappa^{<\kappa} = \kappa$, then $\mathbb{P}_{\mathscr{F},A}$ is $\kappa^+$-c.c. In particular, if $\kappa = \omega$ or $\kappa$ is inaccessible, then $\mathbb{P}_{\mathscr{F},A}$ is $\kappa^+$-c.c.*

*Proof* Note that for $(p, q)$ and $(p, r)$ in $\mathbb{P}_{\mathscr{F},A}$, we have $(p, q \cup r)$ is a common extension of $(p, q)$ and $(p, r)$.                                                                                          $\square$

In particular, given $A \subseteq \omega_1$ and an almost disjoint family $\mathscr{F} = \{x_\alpha \mid \alpha < \omega_1\}$ on $\omega$, we can code $A$ by a real $x$ via forcing over $\mathbb{P}_{\mathscr{F},A}$.

**Definition 2.3** We define that $(p, q) \in \mathbb{P}_{\mathscr{F},A}$ if $p$ is a finite subset of $\omega$, $q$ is a finite subset of $A$ and $(p, q) \leq (p', q') \Leftrightarrow \left(p \supseteq p' \wedge q \supseteq q' \wedge \forall \alpha \in q'(p \cap x_\alpha \subseteq p')\right)$.

If $G$ is $\mathbb{P}_{\mathscr{F},A}$-generic over $V$, then $A = \{\alpha < \omega_1 : |x \cap a_\alpha| < \omega\}$ where we have $x = \bigcup \{p \mid \exists q \, ((p, q) \in G)\}$.

## 2.1.2 The Large Cardinal Notion $0^\sharp$

The large cardinal notion $0^\sharp$ is central to this book. For the theory of $0^\sharp$, I refer to [1, 4, 5]. In this section, I give two different definitions of $0^\sharp$ in $\mathsf{Z}_2$ and show that the theory of $0^\sharp$ can be developed in $\mathsf{Z}_2$. In Sect. 2.4, I will show that the two definitions are equivalent. Finally, I present some statements which are equivalent to $0^\sharp$ over $\mathsf{Z}_3$.

First of all, I introduce the theory of well-founded remarkable EM sets, and then define $0^\sharp$ in $\mathsf{Z}_2$ as the real which codes an unique well-founded remarkable EM set (see [1, 4]).

**Definition 2.4** ([1, 4]) For $\mathscr{M}$ a structure and $X$ a subset of the domain of $\mathscr{M}$ linearly ordered by $<$ (not necessarily a relation of $\mathscr{M}$), $\langle X, < \rangle$ is a set of *indiscernibles* for $\mathscr{M}$ if for every formula $\varphi(v_1, \ldots, v_n)$ in the language of $\mathscr{M}$ with $x_1 < \cdots < x_n$ and $y_1 < \cdots < y_n$ all in $X$, we have

$$\mathscr{M} \models \varphi[x_1, \ldots, x_n] \Leftrightarrow \mathscr{M} \models \varphi[y_1, \ldots, y_n]. \tag{2.1}$$

Note that (2.1) expresses that for $n \in \omega$, all increasing $n$-tuples from $X$ have the same first-order properties in $\mathscr{M}$.

**Definition 2.5** ([1, 4]) Let $\mathfrak{L}_{st}^*$ be $\mathfrak{L}_{st}$ augmented by constants $\{c_k \mid k \in \omega\}$. The theory of the structure $\langle L_\alpha, \in, \gamma_k \rangle_{k \in \omega}$ in $\mathfrak{L}_{st}^*$ is called an EM (for *Ehrenfeucht-Mostowski*) set, where $\alpha$ is a countable limit ordinal $> \omega$ and $\{\gamma_k \mid k \in \omega\}$ is a set of ordinal indiscernibles for $\langle L_\alpha, \in \rangle$ indexed in increasing order.

**Definition 2.6** ([1, 4]) Assume $\Sigma$ is an EM set and $\alpha$ is an infinite countable ordinal. Then $(\mathscr{A}, H)$ is called a $(\Sigma, \alpha)$ *model* if

(a) $\mathscr{A} = \langle A, E \rangle$ is a model of $\mathsf{ZF} + V = L$;
(b) $H \subseteq \mathrm{Ord}^{\mathscr{A}}$ is a set of ordinal indiscernible for $\mathscr{A}$ with order type $\alpha$;
(c) $\mathscr{A} = \mathscr{A} \upharpoonright H$;
(d) $\Sigma$ is a set of $\mathfrak{L}_{st}$-formulas which are valid in $\mathscr{A}$ on increasing sequences from $H$.

**Definition 2.7**  ([1, 4])

(a)  An EM set $\Sigma$ is cofinal if it contains all formulas in the form

$$\text{``}t(v_0, \ldots, v_{n-1}) \in \text{Ord} \to t(v_0, \ldots, v_{n-1}) < v_n\text{''}$$

for any Skolem term $t$.

(b)  An EM set $\Sigma$ is remarkable if for any Skolem term $t$, if the formula

$$\text{``}t(v_0, \ldots, v_{n-1}, v_n, \ldots, v_{n+m}) < v_n\text{''}$$

is in $\Sigma$, then the formula

$$\text{``}t(v_0, \ldots, v_{n-1}, v_n, \ldots, v_{n+m}) = t(v_0, \ldots, v_{n-1}, v_{n+m+1}, \ldots, v_{n+2m+1})\text{''}$$

is in $\Sigma$.

(c)  An EM set $\Sigma$ is well-founded if for any infinite countable ordinal $\alpha$, the $(\Sigma, \alpha)$ model is well-founded.

**Fact 2.1**  ([1, 4]) *Let $\Sigma$ be an EM set. For any infinite countable ordinal $\alpha$, there is an unique (up to isomorphism) $(\Sigma, \alpha)$ model.*

We will be interested in well-founded $(\Sigma, \alpha)$ model. If $\Sigma$ is a well-founded EM set, then for any infinite countable ordinal $\alpha$, there is an unique transitive $(\Sigma, \alpha)$ model and we denote it by $\mathcal{M}(\Sigma, \alpha)$.

**Proposition 2.3**  ($Z_2$) *If there exists a well-founded remarkable cofinal EM set, then it is unique.*

*Proof* Let $\Sigma$ be a well-founded remarkable cofinal EM set. Let $(L_\alpha, H)$ be the unique transitive (up to isomorphism) $(\Sigma, \omega_1^{CK})$ model. Let $(h_\theta : \theta < \omega_1^{CK})$ be an increasing enumeration of $H$. Then $\varphi(v_0, \ldots, v_n) \in \Sigma \Leftrightarrow L_\alpha \models \varphi[h_0, \ldots, h_n]$. Thus, $\Sigma$ is unique.                                                                                           $\square$

In the following, I define the unique well-founded remarkable cofinal EM set, if it exists, as $0^\sharp$. The set $0^\sharp$ is, strictly speaking, a set of formulas. But we can identify $0^\sharp$ with the set of Gödel numbers of sentences in $0^\sharp$ and regard $0^\sharp$ as a subset of $\omega$.

**Proposition 2.4**  ([4]) *If the set $0^\sharp$ exists, then:*

(1)  *$0^\sharp$ is a $\Pi_2^1$ singleton. i.e. $0^\sharp$ is an unique solution of a $\Pi_2^1$ predicate.[1] As a corollary, $0^\sharp$ is a $\Delta_3^1$ real;*

(2)  *Any $x \in \mathscr{P}(\omega) \cap L$ is one-one reducible to $0^\sharp$: there exists an injective total recursive function $f : \omega \to \omega$ such that $x = f^{-1}(0^\sharp)$. I.e. $x \leq_T 0^\sharp$ for any $x \in \omega^\omega \cap L$.*

---

[1] The property $\Sigma = 0^\sharp$ is $\Pi_1$ over $(HC, \in)$, and therefore a $\Pi_2^1$ statement.

**Proposition 2.5** ([1]) *The property "$\Sigma$ is a well-founded remarkable* **EM** *set" is absolute for every inner model of* **ZF**. *Hence* $M \models$ *"$0^{\sharp}$ exists" if and only if $0^{\sharp} \in M$ in which case* $(0^{\sharp})^M = 0^{\sharp}$.

**Definition 2.8** ([5]) Let $M$ be a transitive model of **ZFC** and ($\kappa$ is a cardinal)$^M$. We express that $U$ is an *$M$-ultrafilter on $\kappa$* if $\langle M, \in, U \rangle \models U$ is a normal ultrafilter over $\kappa$:[2]

(i) $U \subseteq \mathscr{P}^M(\kappa)$ is a non-principal ultrafilter on $\mathscr{P}^M(\kappa)$;
(ii) $U$ is $\kappa$-complete: if $\alpha < \kappa$ and $\{X_{\xi} \mid \xi < \alpha\} \in M$ is such that $X_{\xi} \in U$ for any $\xi < \alpha$, then $\bigcap_{\xi < \alpha} X_{\xi} \in U$;
(iii) $U$ is normal: if $f \in M$ is a regressive function on $X \in U$, then $f$ is constant on some $Y \in U$.

We express that $\langle M, \in, U \rangle$ is a **ZFC**$^-$ *premouse* (at $\kappa$) if $U$ is an $M$-ultrafilter on $\kappa$ and $M = L_{\alpha}[U]$ for some $\alpha$ (we stipulate that $L_{\alpha}[U]$ includes the possibility that $\alpha = $ **Ord**, i.e. $M = L[U]$). A premouse $\langle M, \in, U \rangle$ is *iterable* if $U$ is an iterable $M$-ultrafilter. For the definition of "$U$ is an iterable $M$-ultrafilter", I refer to [5]. $U$ is *countably complete* if for $\{X_n : n \in \omega\} \subseteq U$, $\bigcap_{n \in \omega} X_n \neq \emptyset$.

**Fact 2.2** ([5]) *If $U$ is countably complete, then $\langle M, \in, U \rangle$ is iterable.*

In the following, I define that $0^{\sharp}$ exists if there exists a countable iterable premouse. Thus, under this definition, we can identify $0^{\sharp}$ as the real which codes a countable iterable premouse.

Now, we compute the complexity of the statement "$0^{\sharp}$ exists". Recall that any $x \in \omega^{\omega}$ encodes a binary relation $E_x = \{(m, n) \mid x(\langle m, n \rangle) = 0\}$ and consequently a structure $M_x = \langle \omega, E_x \rangle$ in $\mathfrak{L}_{st}$. Let $A_x = \{n \mid x(n) = 1\}$, then $x$ also encodes a structure $N_x = \langle \omega, E_x, A_x \rangle$ in $\mathfrak{L}_{st}(\dot{A})$. If $N_x$ is well-founded and extensional, then it has a transitive collapse $tr(N_x)$.

**Proposition 2.6** ([5]) *The statement "$0^{\sharp}$ exists" is $\Sigma^1_3$.*

*Proof* Let $X = \{x \in \omega \mid N_x$ is well-founded and extensional and $tr(N_x)$ is a premouse$\}$. Define $R \subseteq (\omega^{\omega})^3$ that $R(a, b, z)$ if

(1) $E_a = \{(m, n) \mid a(\langle m, n \rangle) = 0\}$ is a well-ordering with field $\omega$;
(2) $b$ codes a set $\{z_n^b \mid n \in \omega\} \subseteq Z$ together with codes for embeddings $i_{mn}$ : $tr(N_{z_m^b}) \prec tr(N_{z_n^b})$ for $(m, n) \in E_a$ such that:
(3) if $n$ is the minimum in terms of $E_a$, then $z_n^b = z$;
(4) if $n$ is the immediate successor of $m$ in terms of $E_a$, then $tr(N_{z_n^b})$ is the ultrapower of $tr(N_{z_m^b})$ and $i_{mn}$ the corresponding embedding; and
(5) if $n$ is a limit point of $E_a$, then $tr(N_{z_n^b})$ is the direct limit of the corresponding structures and embeddings, and $i_{mn}$ for $(m, n) \in E_a$ is the corresponding embedding into the direct limit.

---

[2]Especially, if $M = V$, Definition 2.8 gives us the definition of a normal measure on cardinals.

Define $S \subseteq (\omega^\omega)^2$ that $S(a, b)$ if and only if (1), (2), (4), and (5) above; (6) if $E_a$ has a maximum element $n$, then the ultrapower of $tr(N_{z_n})$ is well-founded; and (7) if $E_a$ has no maximum element, then the direct limit of the structures and embeddings coded by $b$ is well-founded.

**Fact 2.3** ([5]) $X, R, S$ are all $\Pi_1^1$.

Note that $0^\sharp$ exists if and only if $\exists x(x \in X \wedge tr(N_x)$ is iterable) if and only if $\exists x(x \in X \wedge \forall a \forall b(R(a, b, x) \rightarrow S(a, b)))$ which is a $\Sigma_3^1$ statement.                                  $\square$

**Definition 2.9**  Suppose that $I$ is a set of $<$-indiscernibles over a structure $\mathcal{M}$. Then the indiscernibility type $\Sigma$ of $I$ is defined as the set of all formulas $\varphi(v_1, \ldots, v_n)$ such that $\mathcal{M} \models \varphi[i_1, \ldots, i_n]$ where $i_1, \ldots, i_n \in I$ and $i_1 < \cdots < i_n$.

Finally, I list some statements which are equivalent to "$0^\sharp$ exists" in $\mathsf{Z}_3$.

**Theorem 2.2**  ($\mathsf{Z}_3$, [4–6]) *The following statements are equivalent.*

*(1)  $0^\sharp$ exists.*
*(2)  $L_{\omega_1}$ has an uncountable set of indiscernibles.*
*(3)  There exists an uncountable subset $C \subseteq \omega_1$ such that for any formula $\varphi$ and for any two increasing sequences $\xi_0 < \cdots < \xi_{n-1}$ and $\xi_0' < \cdots < \xi_{n-1}'$ of elements from $C$, we have*

$$L_{\omega_1} \models \varphi[\xi_0, \ldots, \xi_{n-1}] \leftrightarrow L_{\omega_1} \models \varphi[\xi_0', \ldots, \xi_{n-1}'].$$

*(4)  For each formula $\varphi$, there exists a closed unbounded subset $C$ of $\omega_1$ such that either*

    *(a)  $L_{\omega_1} \models \varphi[\xi_0, \ldots, \xi_{n-1}]$ for any increasing sequence $\xi_0 < \cdots < \xi_{n-1}$ of elements from $C$, or*
    *(b)  $L_{\omega_1} \models \neg\varphi[\xi_0, \ldots, \xi_{n-1}]$ for any increasing sequence $\xi_0 < \cdots < \xi_{n-1}$ of elements from $C$.*

*(5)  There exists a set of formulas $\Sigma$ in $\mathfrak{L}_{st}$ such that for every $\alpha < \omega_1$ there exists a set $I_\alpha$ of indiscernibles for $L_{\omega_1}$ such that $o.t.(I_\alpha) = \alpha$ and*

$$\Sigma = \{\varphi(v_1, \ldots, v_n) : L_{\omega_1} \models \varphi[i_1, \ldots, i_n]\}$$

*where $i_1, \ldots, i_n \in I_\alpha$ and $i_1 < \cdots < i_n$.*
*(6)  There exists a well-founded, cofinal and remarkable* **EM** *set.*

*Remark 2.1*  In $\mathsf{Z}_3$, we can show that $0^\sharp$ exists if and only if $L_{\omega_1}$ has an uncountable set of indiscernibles (see [4, 6]). In the proof of this theorem, the existence of uncountably many indiscernibles is really not required to show the existence of $0^\sharp$. It suffices to know only that sets of indiscernibles of every order type $\alpha < \omega_1$ can be found, all of which have the same indiscernibility type over the given structure.

### 2.1.3 Remarkable Cardinal

The notion of *remarkable cardinal* is another central notion of large cardinals in this book. The notion of remarkable cardinal is first introduced by Ralf Schindler in [7] to show that the statement "the theory of $L(\mathbb{R})$ cannot be changed by proper forcing" is equiconsistent with the existence of a remarkable cardinal.

**Definition 2.10** (Schindler, [7, 8]) A cardinal $\kappa$ is remarkable if for every regular cardinal $\lambda > \kappa$, there is a regular cardinal $\overline{\lambda} < \kappa$ such that in $V^{Col(\omega, <\kappa)}$ there is an elementary embedding $j : H_{\overline{\lambda}}^V \to H_\lambda^V$ with $j(crit(j)) = \kappa$.

We can view a remarkable cardinal as a type of generic supercompact cardinal using the following theorem of Magidor.

**Theorem 2.3** (Magidor, [9]) *A cardinal $\kappa$ is supercompact if and only if for every regular cardinal $\lambda > \kappa$ there is a regular cardinal $\overline{\lambda} < \kappa$ and an elementary embedding $j : H_{\overline{\lambda}} \to H_\lambda$ with $j(crit(j)) = \kappa$.*

Remarkable cardinal has several equivalent formulations. In this book, I use the following formulation of remarkable cardinal due to Schindler from [10] which is equivalent to Definition 2.10.

**Definition 2.11** ([10])

(1) Let $\kappa$ be a cardinal. Let $G$ be $Col(\omega, <\kappa)$-generic over $V$, let $\theta > \kappa$ be a regular cardinal and let $X \in [H_\theta^{V[G]}]^\omega$. We say that $X$ condense remarkably if $X = ran(\pi)$ for some elementary $\pi : (H_\beta^{V[G \cap H_\alpha^V]}, \in, H_\beta^V, G \cap H_\alpha^V) \to (H_\theta^{V[G]}, \in, H_\theta^V, G)$ where $\alpha = crit(\pi) < \beta < \kappa$ and $\beta$ is a regular cardinal in $V$.

(2) A cardinal $\kappa$ is remarkable if for all regular cardinal $\theta > \kappa$ we have $\Vdash_{Col(\omega, <\kappa)}^V$ "$\{X \in [H_{\check{\theta}}^{V[\dot{G}]}]^\omega \cap V[\dot{G}] : X$ condense remarkably$\}$ is stationary".

The following fact summarizes the strength of remarkable cardinal in the hierarchy of large cardinals. For the definition of totally indescribable cardinal, ineffable cardinal, iterable cardinal and Edrös cardinal, I refer to Appendix C.

**Fact 2.4** ([7, 10, 11])

*(1) Every remarkable cardinal is remarkable in L.*

*(2) Every remarkable cardinal $\kappa$ is totally indescribable and n-ineffable for every $n < \omega$.*

*(3) If $\kappa$ is a remarkable cardinal, then there is a countable transitive model of* ZFC *with a proper class of 1-iterable cardinals.*

*(4) If $\kappa$ is 2-iterable, then $\kappa$ is a limit of remarkable cardinal.*

*(5) If there exists a $\omega$-Edrös cardinal, then for any $n \in \omega$ there exist $\alpha < \beta < \omega_1$ such that $L_\beta \models$ "ZFC $+ \alpha$ is remarkable".*

*(6) If $0^\sharp$ exists, then every Silver indiscernible is remarkable in L.*

**Lemma 2.2** *Suppose $\kappa$ is an $L$-cardinal. The following are equivalent:*

(a) *$\kappa$ is remarkable in $L$;*
(b) *If $\gamma \geq \kappa$ is an $L$-cardinal, $\theta > \gamma$ is a regular cardinal in $L$, then $\Vdash^L_{Col(\omega, <\kappa)}$*
    *"$\{X : X \prec L_{\check{\theta}}[\dot{G}], |X| = \omega, \check{\gamma} \in X,$ and $\overline{\gamma}$ is an $L$-cardinal\} is stationary";*
(c) *If $\gamma \geq \kappa$ is an $L$-cardinal, $\theta > \gamma$ is a regular cardinal in $L$, then $\Vdash^L_{Col(\omega, <\kappa)}$*
    *"$\{X : X \prec L_{\check{\theta}}[\dot{G}], |X| = \omega$ and $o.t.(X \cap \check{\gamma})$ is an $L$-cardinal\} is stationary".*

*Proof* We only show that $(a) \Leftrightarrow (b)$. The argument for $(a) \Leftrightarrow (c)$ is similar. From Definition 2.11, $\kappa$ is remarkable in $L$ iff if $\theta > \kappa$ is a regular cardinal in $L$ and $G$ is $Col(\omega, <\kappa)$-generic over $L$, then $L[G] \models$ "$\{X \in [L_\theta[G]]^\omega : X = ran(\pi)$ for some elementary $\pi : (L_\beta[G \upharpoonright \alpha], \in, L_\beta, G \upharpoonright \alpha) \to (L_\theta[G], \in, L_\theta, G)$ where $\alpha = crit(\pi) < \beta < \kappa$ and $\beta$ is a regular cardinal in $L\}$ is stationary" iff if $\gamma \geq \kappa$ is an $L$-cardinal, $\theta > \gamma$ is a regular cardinal in $L$ and $G$ is $Col(\omega, <\kappa)$-generic over $L$, then $L[G] \models \{X : X \prec L_\theta, |X| = \omega, \gamma \in X$ and $\overline{\gamma}$ is an $L$-cardinal\} is stationary.                                                                    □

## 2.2   The Strength of $Z_2$ + Harrington's Principle

In this section, we prove that $Z_2 + HP$ is equiconsistent with ZFC via class forcing. As a corollary, $Z_2 + HP$ does not imply that $0^\sharp$ exists. For the theory of class forcing, I refer to [12]. Our class forcing notions will always be definable ones. The history of the main result in this section is as follows: "$Z_2 + HP$ does not imply that $0^\sharp$ exists" was first proved in [13] via set forcing; however, the large cardinal strength of "$Z_2 + HP$" is not discussed in [13]; in the joint work with Schindler, we compute the exact large cardinal strength of $Z_2 + HP$ in [14]. In Theorem 2.4, we force a class model of $Z_2 + HP$ via class forcing.

**Theorem 2.4** *$Z_2 + HP$ is equiconsistent with ZFC.*

*Proof* It is easy to see that $Z_2 + HP$ implies that $L \models$ ZFC. We now show that $Con(ZFC)$ implies $Con(Z_2 + HP)$. We assume that $L$ is a minimal model of ZFC, i.e.

$$\text{there is no } \alpha \text{ such that } L_\alpha \models \text{ZFC.} \tag{2.2}$$

Let $G$ be $Col(\omega, < Ord)$-generic over $L$. Then $L[G] \models Z_2$. In $L[G]$, we could pick $A \subseteq Ord$ such that $V = L[A]$ and if $\lambda \geq \omega$ is an $L$-cardinal, then $A \cap [\lambda, \lambda + \omega)$ codes a well-ordering of $(\lambda^+)^L$. By (2.2), we will then have that for all $\alpha \geq \omega$,

$$L_{\alpha+1}[A \cap \alpha] \models \alpha \text{ is countable.} \tag{2.3}$$

By (2.3), there exists a canonical sequence $(c_\alpha : \alpha \in Ord)$ of pairwise almost disjoint subset of $\omega$ such that $c_\alpha$ is the $L_{\alpha+1}[A \cap \alpha]$-least subset of $\omega$ and $c_\alpha$ is almost disjoint from every member of $\{c_\beta : \beta < \alpha\}$. Do almost disjoint forcing to code $A$ by a real

$x$ such that for any $\alpha \in \mathrm{Ord}$, we have $\alpha \in A \Leftrightarrow |x \cap c_\alpha| < \omega$. This forcing is *c.c.c.* Note that $L[A][x] = L[x]$ and $L[x] \models Z_2$.

Now we show that HP holds in $L[x]$. It suffices to show that if $\alpha$ is $x$-admissible, then $\alpha$ is an $L$-cardinal. Suppose $\alpha$ is $x$-admissible but is not an $L$-cardinal. Let $\lambda$ be the largest $L$-cardinal $< \alpha$. Note that we can define $A \cap \alpha$ over $L_\alpha[x]$. Since $A \cap [\lambda, \lambda + \omega) \in L_\alpha[x]$ and $A \cap [\lambda, \lambda + \omega)$ codes a well-ordering of $(\lambda^+)^L$, we have $(\lambda^+)^L \in L_\alpha[x]$ since $\alpha$ is $x$-admissible. But $(\lambda^+)^L > \alpha$ which leads to a contradiction. Thus, $L[x] \models Z_2 + \mathrm{HP}$. $\qquad\square$

As a corollary, $Z_2 + \mathrm{HP}$ does not imply that $0^\sharp$ exists. However, from Theorem 1.25, we have $Z_2 + 0^\sharp$ exists implies HP. In fact, $Z_2 + 0^\sharp$ exists implies that if $\alpha$ is $0^\sharp$-admissible, then $\alpha$ can be any large cardinal compatible with $L$ since Silver indiscernibles can have any large cardinal property compatible with $L$.[3]

## 2.3 The Strength of $Z_3$ + Harrington's Principle

In this section, we prove via class forcing that $Z_3 + \mathrm{HP}$ is equiconsistent with ZFC + there is a remarkable cardinal. As a corollary, $Z_3 + \mathrm{HP}$ does not imply that $0^\sharp$ exists. The history of the main result in this section is as follows: "$Z_3 + \mathrm{HP}$ does not imply that $0^\sharp$ exists" was first proved in [13] via set forcing without the use of the reshaping technique; however, the large cardinal strength of $Z_3 + \mathrm{HP}$ is not discussed in [13]; in the joint work with Ralf Schindler, we compute the exact large cardinal strength of $Z_3 + \mathrm{HP}$ in [14]. In Theorem 2.5, assuming there is one remarkable cardinal, we force a class model of $Z_3 + \mathrm{HP}$ via class forcing using the reshaping technique.

**Proposition 2.7** $Z_3 + \mathrm{HP}$ *implies* $L \models \mathrm{ZFC} + \omega_1^V$ *is remarkable.*

*Proof* We assume $Z_3 + \mathrm{HP}$. It is easy to show that $L \models \mathrm{ZFC}$. We now show that $\omega_1^V$ is remarkable in $L$. Suppose $L \models \theta > \omega_1^V$ is regular, and define $\eta = (\theta^+)^L$. Let $x \in \omega^\omega$ witness HP. Let $G$ be $Col(\omega, < \omega_1^V)$-generic over $V$. Let $f : [L_\theta[G]]^{<\omega} \to L_\theta[G]$, $f \in L[G]$ and let $X \prec L_\eta[x][G]$ be such that $|X| = \omega$ and $\{\omega_1, \theta, f\} \subseteq X$. Let $\tau : L_{\bar\eta}[x][G \cap L_\alpha[x]] \cong X$ be the collapsing map, where $\alpha = crit(\tau)$, $\tau(\alpha) = \omega_1^V$ and $\tau(\bar f) = f$. Since $\bar\eta$ is $x$-admissible, $\bar\eta$ is an $L$-cardinal by the choice of $x$ as witnessing HP, and thus $\beta = o.t.(X \cap \theta) = \tau^{-1}(\theta)$ is a regular $L$-cardinal. Thus, $X \cap L_\theta[G]$ condenses remarkably. By absoluteness, there is in $L[G]$ some elementary $\overline\tau : L_{\bar\eta}[G \cap L_\alpha] \to L_\eta[G]$ such that $\overline\tau(\beta) = \theta$ and $\overline\tau(\bar f) = f$. That is, in $L[G]$, there is some $X \in [H_\theta^{L[G]}]^\omega \cap L[G]$ which condenses remarkably and is closed under $f$. Hence, by Definition 2.11, $\omega_1^V$ is remarkable in $L$. $\qquad\square$

---

[3]Examples of notions of large cardinals compatible with $L$ are: inaccessible cardinal, reflecting cardinal, Mahlo cardinal, weakly compact, indescribable cardinal, unfoldable cardinal, subtle cardinal, ineffable cardinal, 1-iterable cardinal, remarkable cardinal, 2-iterable cardinal and $\omega$-Erdös cardinal. For definitions of these large cardinal notions, I refer to Sect. 2.1.3 and Appendix C.

**Proposition 2.8** *If "ZFC+ there exists a remarkable cardinal" is consistent, then "Z$_3$ + HP" is consistent.*

*Proof* We assume that $L \models$ "ZFC $+ \kappa$ is a remarkable cardinal" and

$$\text{there is no } \alpha \text{ such that } L_\alpha \models \text{``ZFC} + \kappa \text{ is a remarkable cardinal''}. \tag{2.4}$$

In the following, we write $S_\mu$ for $\{X \in [L_\mu]^\omega : X \prec L_\mu$ and $o.t.(X \cap \mu)$ is an $L$-cardinal$\}$, as defined in the respective models of set theory which are to be considered.

From Lemma 2.2, $\kappa$ is remarkable in $L$ if and only if for any $L$-cardinal $\mu \geq \kappa$, for any $G$ which is $Col(\omega, < \kappa)$-generic over $L$, $L[G] \models S_\mu$ is stationary. Let $G$ be $Col(\omega, < \kappa)$-generic over $L$. Since $\kappa$ is remarkable in $L$, $L[G] \models$ "$S_\mu$ is stationary for any $L$-cardinal $\mu \geq \kappa$". Let $H$ be $Col(\kappa, < \text{Ord})$-generic over $L[G]$. Note that $Col(\kappa, < \text{Ord})$ is countably closed. Standard arguments give that

$$L[G][H] \models \text{``Z}_3 + S_\mu \text{ is stationary for all L-cardinals } \mu > \kappa\text{''}. \tag{2.5}$$

In $L[G][H]$, we may pick some $B \subseteq \text{Ord}$ such that $V = L[B]$ and if $\lambda \geq \omega_1$ is an $L$-cardinal, then $B \cap [\lambda, \lambda + \omega_1)$ codes a well-ordering of $(\lambda^+)^L$. By (2.4), we have that for all $\alpha \geq \omega_1$,

$$L_{\alpha+1}[B \cap \alpha] \models |\alpha| \leq \aleph_1. \tag{2.6}$$

By (2.6), there exists then a canonical sequence $(C_\alpha | \alpha \in \text{Ord})$ of pairwise almost disjoint subsets of $\omega_1$ such that $C_\alpha$ is the $L_{\alpha+1}[B \cap \alpha]$-least subset of $\omega_1$ and $C_\alpha$ is almost disjoint from every member of $\{C_\beta | \beta < \alpha\}$. Do almost disjoint forcing to code $B$ by some $A \subset \omega_1$ such that for any $\alpha \in \text{Ord}$, we have $\alpha \in B \Leftrightarrow |A \cap C_\alpha| < \omega_1$. This forcing is countably closed and has the Ord-c.c. Note that $L[B][A] = L[A]$, $L[A] \models Z_3$ and

$$L[A] \models S_\mu \text{ is stationary for any } L\text{-cardinal } \mu \geq \kappa. \tag{2.7}$$

Suppose $\alpha > \omega_1$ is $A$-admissible, but $\alpha$ is not an $L$-cardinal. Let $\lambda$ be the largest $L$-cardinal $< \alpha$. Note that $\lambda + \omega_1 < \alpha$ and we can compute $B \cap \alpha$ over $L_\alpha[A]$. Hence $B \cap [\lambda, \lambda + \omega_1) \in L_\alpha[A]$, and $B \cap [\lambda, \lambda + \omega_1)$ codes a well-ordering of $(\lambda^+)^L$. Thus, $(\lambda^+)^L < \alpha$ since $\alpha$ is $A$-admissible, which leads to a contradiction. We have shown that in $L[A]$,

$$\text{every A-admissible ordinal above } \omega_1 \text{ is an } L\text{-cardinal}. \tag{2.8}$$

Now over $L[A]$ we do reshaping as follows (cf. [15, Sect. 1.3] on the original reshaping forcing).

**Definition 2.12** Define that $p \in \mathbb{P}$ if and only if $p : \alpha \to 2$ for some $\alpha < \omega_1$ and for any $\xi \leq \alpha$, there is $\gamma$ such that $L_\gamma[A \cap \xi, p \upharpoonright \xi] \models$ "$\xi$ is countable" and every $(A \cap \xi)$-admissible $\lambda \in [\xi, \gamma]$ is an $L$-cardinal.

It is easy to check the extendability property of $\mathbb{P}$: $(\forall p \in \mathbb{P})(\forall \alpha < \omega_1)(\exists q \leq p)(dom(q) \geq \alpha)$. Note that $|\mathbb{P}| = \omega_1$ since CH holds true in $L[A]$.

**Lemma 2.3** $\mathbb{P}$ is $\omega$-distributive.[4]

*Proof* Our argument is a variant of an argument from [16] (cf. also [17]). Let $p \in \mathbb{P}$ and $\overrightarrow{D} = (D_n : n \in \omega)$ be a sequence of open dense sets. Take $\nu > \omega_1$ such that $\overrightarrow{D} \in L_\nu[A]$ and $L_\nu[A]$ is a model of a reasonable fragment of $\mathsf{ZFC}^-$. By (2.8) we have that

$$L_\mu[A] \models \text{"any } A\text{-admissible ordinal} \geq \omega_1 \text{ is an } L\text{-cardinal"}, \qquad (2.9)$$

where $\mu = (\nu^+)^L$. By (2.7) we can pick $X$ such that $\pi : L_{\bar\mu}[A \cap \delta] \cong X \prec L_\mu[A]$, $|X| = \omega$, $\{p, \mathbb{P}, A, \overrightarrow{D}, \omega_1, \nu\} \subseteq X$, $\overline{\mu}$ is an $L$-cardinal, $\pi(\delta) = \omega_1$ and $\delta = crit(\pi)$. Note that (2.9) yields that $L_{\bar\mu}[A \cap \delta] \models \text{"every } A \cap \delta\text{-admissible ordinal} \geq \delta \text{ is an } L\text{-cardinal"}$. Since $\overline{\mu}$ is an $L$-cardinal, we have that

$$\text{every } A \cap \delta\text{-admissible } \lambda \in [\delta, \overline{\mu}] \text{ is an } L\text{-cardinal.} \qquad (2.10)$$

Let $\pi(\overline{\nu}) = \nu$, $\pi(\overline{\mathbb{P}}) = \mathbb{P}$ and $\pi(\overline{D}) = \overrightarrow{D}$ with $\overline{D} = (\overline{D}_n : n \in \omega)$. By (2.6), we may let $(E_i : i < \delta) \in L_{\bar\mu}[A \cap \delta]$ be an enumeration of all clubs in $\delta$ which exist in $L_{\bar\nu}[A \cap \delta]$. Let $E$ be the diagonal intersection of $(E_i : i < \delta)$. Note that $E \setminus E_i$ is bounded in $\delta$ for all $i < \delta$. In $L[A]$, let us pick a strictly increasing sequence $(\varepsilon_n : n < \omega)$ such that $\{\varepsilon_n : n < \omega\} \subseteq E$ and $(\varepsilon_n : n < \omega)$ is cofinal in $\delta$.

We want to find a $q \in \mathbb{P}$ such that $q \leq p$, $dom(q) = \delta$, $L_{\bar\mu}[A \cap \delta, q] \models \text{"}\delta$ is countable" and $q \in \overline{D}_n$ for all $n \in \omega$. For this we construct a sequence $(p_n : n \in \omega)$ of conditions such that $p_0 = p$, $p_{n+1} \leq p_n$ and $p_{n+1} \in \overline{D}_n = D_n \cap L_{\bar\nu}[A \cap \delta]$ for all $n \in \omega$. Also we construct a sequence $\{\delta_n : n \in \omega\}$ of ordinals. Suppose $p_n \in L_{\bar\nu}[A \cap \delta]$ is given. Let $\gamma = dom(p_n)$. Note that $\gamma < \delta$ since $p_n \in L_{\bar\nu}[A \cap \delta]$. Now we work in $L_{\bar\nu}[A \cap \delta]$. By extendability, for all $\xi$ with $\gamma \leq \xi < \delta$ we may pick some $p^\xi \leq p_n$ such that $p^\xi \in \overline{D}_n$, $dom(p^\xi) > \xi$ and for all limit ordinal $\lambda$ with $\gamma \leq \lambda \leq \xi$ we have $p^\xi(\lambda) = 1$ if and only if $\lambda = \xi$. There exists $C \in L_{\bar\nu}[A \cap \delta]$ which is a club on $\delta$ such that for any $\eta \in C$, $\xi < \eta$ implies $dom(p^\xi) < \eta$.

Next, we work in $L_{\bar\mu}[A \cap \delta]$. We may pick some $\eta \in E$ such that $\varepsilon_n \leq \eta$ and $E \setminus C \subseteq \eta$. Let $p_{n+1} = p^\eta$ and $\delta_n = \eta$. Note that $p_{n+1} \leq p_n$ and $p_{n+1} \in \overline{D}_n$. Also $dom(p_{n+1}) < min(E \setminus (\delta_n + 1))$ so that for all limit ordinal $\lambda \in E \cap (dom(p_{n+1}) \setminus dom(p_n))$, we have $p_{n+1}(\lambda) = 1$ if and only if $\lambda = \delta_n$.

Now let $q = \bigcup_{n \in \omega} p_n$. We need to check that $q \in \mathbb{P}$. Note that $dom(q) = \delta$. By (2.10), it suffices to check that $L_{\bar\mu}[A \cap \delta, q] \models \delta$ is countable. From the construction of $p_n$'s, we have $\{\lambda \in E \cap (dom(q) \setminus dom(p)) : \lambda \text{ is a limit ordinal and } q(\lambda) = 1\} = \{\delta_n : n \in \omega\}$, which is cofinal in $\delta$ since $\delta_n \geq \varepsilon_n$ for all $n < \omega$. Recall that $E \in L_{\bar\mu}[A \cap \delta, q]$. Thus, $\{\delta_n : n \in \omega\} \in L_{\bar\mu}[A \cap \delta, q]$ witnesses that $\delta$ is countable in $L_{\bar\mu}[A \cap \delta, q]$. $\square$

---

[4]Recall that $\mathbb{P}$ is $\omega$-distributive if every function $f : \alpha \to V$ in the generic extension with $\alpha < \omega_1$ is in the ground model.

The proof of Lemma 2.3 can be adapted to show that $\mathbb{P}$ is stationary preserving (cf. [17]), namely as follows.

Forcing with $\mathbb{P}$ adds some $F : \omega_1 \to 2$ such that for any $\alpha < \omega_1$ there exists $\gamma$ such that $L_\gamma[A \cap \alpha, F \restriction \alpha] \models$ "$\alpha$ is countable" and every $(A \cap \alpha)$-admissible $\lambda \in [\alpha, \gamma]$ is an $L$-cardinal; for each $\alpha < \omega_1$ let $\alpha^*$ be the least such $\gamma$. Let $D = A \oplus F$. We may assume that for any $L$-cardinal $\lambda < \omega_1^V$, $D$ restricted to odd ordinals in $[\lambda, \lambda + \omega)$ codes a well-ordering of the least $L$-cardinal $> \lambda$. By Lemma 2.3, we have $L[A][F] = L[D] \models Z_3$.

Now we do almost disjoint forcing over $L[D]$ to code $D$ by a real $x$. There exists a canonical sequence $(x_\alpha | \alpha < \omega_1)$ of pairwise almost disjoint subset of $\omega$ such that $x_\alpha$ is the $L_{\alpha^*}[D \cap \alpha]$-least subset of $\omega$ such that $x_\alpha$ is almost disjoint from every member of $\{x_\beta | \beta < \alpha\}$. Almost disjoint forcing adds a real $x$ such that for all $\alpha < \omega_1$, $\alpha \in D \Leftrightarrow |x_\alpha \cap x| < \omega$. The forcing has the c.c.c., and thus $L[D][x] = L[x] \models Z_3$.

We finally claim that $L[x] \models$ HP. Suppose $\alpha$ is $x$-admissible. We show that $\alpha$ is an $L$-cardinal. If $\alpha \geq \omega_1$, then $\alpha$ is also $A$-admissible and hence is an $L$-cardinal by (2.8). Now we assume that $\alpha < \omega_1$ and $\alpha$ is not an $L$-cardinal. Let $\lambda$ be the largest $L$-cardinal $< \alpha$. Recall that for $\xi < \omega_1, \xi^* > \xi$ is least such that $L_{\xi^*}[A \cap \xi, F \restriction \xi] \models \xi$ is countable. Every $(D \cap \xi)$-admissible $\lambda' \in [\xi, \xi^*]$ is an L-cardinal.

**Case 1** For all $\xi < \lambda + \omega, \xi^* < \alpha$. Then $D \cap (\lambda + \omega)$ can be computed inside $L_\alpha[x]$. But then, since $\alpha$ is $x$-admissible, the ordinal coded by $D$ restricted to the odd ordinals in $[\lambda, \lambda + \omega)$, namely the least $L$-cardinal $> \lambda$, is in $L_\alpha[x]$ and hence $(\lambda^+)^L < \alpha$, which leads to a contradiction.

**Case 2** Not Case 1. Let $\xi < \lambda + \omega$ be least such that $\xi^* \geq \alpha$. Then $D \cap \xi$ can be computed inside $L_\alpha[x]$. As $\alpha$ is $x$-admissible, we have $\alpha$ is $(D \cap \xi)$-admissible. But all $(D \cap \xi)$-admissibles $\lambda' \in [\xi, \xi^*]$ are $L$-cardinals, so that $\alpha$ is an $L$-cardinal by $\xi < \alpha \leq \xi^*$, which leads to a contradiction. We have shown that $L[x] \models Z_3 + \text{HP}$.  $\square$

**Theorem 2.5** *The following two theories are equiconsistent:*

*(1)* $Z_3 + \text{HP}$.
*(2)* ZFC+ *there exists a remarkable cardinal.*

*Proof* Follows from Propositions 2.7 and 2.8.  $\square$

## 2.4   $Z_4$ + Harrington's Principle Implies that $0^\sharp$ Exists

In this section, I prove in $Z_4$ that HP implies that $0^\sharp$ exists. From Sects. 2.2 and 2.3, "HP implies that $0^\sharp$ exists" is neither provable in $Z_2$ nor provable in $Z_3$. As a corollary, $Z_4$ is the minimal system in higher-order arithmetic for proving that HP implies that $0^\sharp$ exists. Finally, I introduce generalized Harrington's Principle $\text{HP}(M)$ and give particular characterizations of $\text{HP}(M)$ for various known examples of inner models $M$. I also show that in some cases, this generalized principle fails.

In Sect. 2.1.2, I give two different definitions of $0^\sharp$ in $Z_2$: (1) $0^\sharp$ is the unique well-founded remarkable cofinal EM set; (2) $0^\sharp$ is the real which codes a countable iterable premouse. In this section, I prove in $Z_4$ that HP implies that $0^\sharp$ exists via these two definitions.

First of all, we prove in $Z_4$ that HP implies that $0^\sharp$ exists via defining $0^\sharp$ as the unique well-founded remarkable cofinal EM set.

**Theorem 2.6** ($Z_4$) HP *implies that* $L_{\omega_2}$ *has an uncountable set of indiscernibles.*

*Proof* Let $a$ be the witness real for HP. We work in $L[a]$. Pick $\eta > \omega_2$ and $N$ such that $\eta$ is $a$-admissible, $N \prec L_\eta[a]$, $\omega_2 \in N$, $|N| = \omega_1$ and $N$ is closed under $\omega$-sequences. Let $j : L_\theta[a] \cong N \prec L_\eta[a]$ be the inverse of the collapsing map and $\kappa = crit(j)$. By HP, we have $\theta$ is an $L$-cardinal. Define $U = \{X \subseteq \kappa \mid X \in L \wedge \kappa \in j(X)\}$. Note that $(\kappa^+)^L \leq \theta < \omega_2$ and $U \subseteq L_\theta$ is an $L$-ultrafilter on $\kappa$.

Do the ultrapower construction for $\langle L_{\omega_2}, \in, U \rangle$ as follows. Consider the structure $\langle L_{\omega_2}, \in, U \rangle$. For $f, g : \kappa \to L_{\omega_2}$ with $f, g \in L_{\omega_2}$, define $f \sim g$ iff $\{\alpha < \kappa : f(\alpha) = g(\alpha)\} \in U$. For $f : \kappa \to L_{\omega_2}$ with $f \in L_{\omega_2}$, define $[f] = \{g : \kappa \to L_{\omega_2} \mid g \in L_{\omega_2}, g \sim f$ and for any $h : \kappa \to L_{\omega_2}$ with $h \in L_{\omega_2}$ (if $h \sim f$, then $rank(g) \leq rank(h))\}$. Let $L_{\omega_2}/U = \{[f] : f \in L_{\omega_2}$ and $f : \kappa \to L_{\omega_2}\}$.

Then $L_{\omega_2} \prec L_{\omega_2}/U$ via the map which sends $x \in L_{\omega_2}$ to $[c_x]$ where $c_x : \kappa \to \{x\}$ is the constant function with value $x$.

**Claim** $L_{\omega_2}/U$ *is well-founded.*

*Proof* If not, then there is a sequence $\langle f_n : n \in \omega \rangle$ of functions such that for any $n \in \omega$, $f_n : \kappa \to L_{\omega_2}$, $f_n \in L_{\omega_2}$ and $U_n = \{\alpha \in \kappa : f_{n+1}(\alpha) \in f_n(\alpha)\} \in U$. Since $\mathcal{P}(\kappa) \cap L \subseteq L_\theta$, the sequence $\langle U_n : n \in \omega \rangle$ is a sequence of elements of $L_\theta$. Since $M$ is closed under $\omega$-sequence from $M$, we have $L_\theta[a]$ is closed under $\omega$-sequence from $L_\theta[a]$. Hence $\langle U_n : n \in \omega \rangle \in L_\theta[a]$. Since $\kappa \in j(U_n)$ for any $n \in \omega$ and $j$ is elementary, we have $\kappa \in j(\bigcap_{n \in \omega} U_n)$. Thus, $\bigcap_{n \in \omega} U_n \neq \emptyset$. Take $\xi \in \bigcap_{n \in \omega} U_n$. Then $f_{n+1}(\xi) \in f_n(\xi)$ for any $n \in \omega$, which leads to a contradiction. □

Since $L_{\omega_2} \models V = L$, $L_{\omega_2} \prec L_{\omega_2}/U$ and $L_{\omega_2}/U$ is well-founded, we have $L_{\omega_2}/U \cong L_\beta$ for some $\beta$. Since $|L_{\omega_2}/U| \leq \omega_2$, we have $\beta \leq \omega_2$. Thus, $\beta = \omega_2$ and we get a nontrivial elementary embedding $e : L_{\omega_2} \prec L_{\omega_2}$ with $crit(e) = \kappa$.

**Lemma 2.4** HP *implies that there exists a club on* $\omega_2$ *of regular $L$-cardinals.*

*Proof* Now we show that there exists a club on $\omega_2$ of regular $L$-cardinals. Suppose $X \prec L_\eta[a]$, $\omega_1 \subseteq X$ and $\omega_2 \in X$. The transitive collapse of $X$ is $L_{\bar\eta}[a]$ for some $\bar\eta$. Since $L_\eta \models \omega_2$ is a regular cardinal, we have $L_{\bar\eta} \models \overline{\omega_2}$ is a regular cardinal. By HP, $\bar\eta$ is an $L$-cardinal and hence $\overline{\omega_2}$ is a regular $L$-cardinal. Since $\omega_1 \subseteq X$, we have $\overline{\omega_2} = X \cap \omega_2$. We have shown that if $X \prec L_\eta[a]$, $\omega_1 \subseteq X$ and $\omega_2 \in X$, then $X \cap \omega_2 = \overline{\omega_2}$ is a regular $L$-cardinal. Thus, there exists a club on $\omega_2$ of regular $L$-cardinals. □

Let $D$ be a club on $\omega_2$ of regular $L$-cardinals such that $D \cap (\kappa + 1) = \emptyset$.

**Lemma 2.5** *For any $\alpha \in D$, we have $e(\alpha) = \alpha$.*

*Proof* Suppose $\alpha \in D$ and $f \in L_{\omega_2}$ where $f : \kappa \to \alpha$. Since $\alpha > \kappa$ is a regular $L$-cardinal, $f$ is bounded by some $\eta < \alpha$. Thus, $[f] < [c_\eta]$. Hence $e(\alpha) = \lim_{\beta \to \alpha} e(\beta)$. If $\beta < \alpha$, then $|e(\beta)| \le (|\beta^\kappa|)^L \le \alpha$. Thus, $e(\alpha) = \alpha$. $\square$

We define a sequence $\langle C_\alpha : \alpha < \omega_1 \rangle$ as follows. Let $C_0 = D$. For any $\nu < \omega_1$, $C_{\nu+1} = \{\mu \in C_\nu : \mu$ is the $\mu$-th element of $C_\nu$ in the increasing enumeration of $C_\nu\}$. If $\nu \le \omega_1$ is a limit ordinal, let $C_\nu = \bigcap_{\beta < \nu} C_\beta$.

**Lemma 2.6** *For $\nu < \omega_1$, if $C_\nu$ is a club on $\omega_2$, then $C_{\nu+1}$ is also a club on $\omega_2$.*

*Proof* Let $f : \omega_2 \to C_\nu$ be an increasing enumeration of $C_\nu$. Note that since $C_\nu$ is a club on $\omega_2$, $f$ is a normal function (increasing and continuous) on $\omega_2$. It is a standard fact that the set of fixed points of a normal function on $\omega_2$ is a club on $\omega_2$. Note that $C_{\nu+1}$ is the set of fixed points of $f$ in $C_\nu$. Thus, $C_{\nu+1}$ is a club on $\omega_2$. $\square$

By induction on $\nu \le \omega_1$, it is easy to check that $C_\nu$ is a club on $\omega_2$. By Lemma 2.5, for $\nu \le \omega_1$, we have $e \upharpoonright C_\nu = id$.

Now we will find $\omega_1$-many indiscernibles for $(L_{\omega_2}, \in)$. The rest of the argument essentially follows from [1, Theorem 18.20]. However, the proof of [1, Theorem 18.20] is not done in $\mathbf{Z}_4$. In the following, I give the proof of [1, Theorem 18.20] in $\mathbf{Z}_4$.

For each $\nu < \omega_1$, let $M_\nu$ be the Skolem hull of $\kappa \cup C_\nu$ in $L_{\omega_2}$. The transitive collapse of $M_\nu$ is $L_{\omega_2}$. Let $i_\nu : L_{\omega_2} \cong M_\nu \prec L_{\omega_2}$ be the inverse of the collapsing map and $\kappa_\nu = i_\nu(\kappa)$.

**Proposition 2.9** *Let $\nu, \tau < \omega_1$. Then*

*(1) $\kappa_\nu$ is the least ordinal in $M_\nu \setminus (\kappa + 1)$;*
*(2) if $\nu < \tau$ and $x \in M_\tau$, then $i_\nu(x) = x$;*
*(3) if $\nu < \tau$, then $i_\nu(\kappa_\tau) = \kappa_\tau$;*
*(4) if $\nu < \tau$, then $\kappa_\nu < \kappa_\tau$.*

*Proof* (1) Since $\kappa \subseteq M_\nu$, we have $i_\nu \upharpoonright \kappa = id$ and $i_\nu(\kappa)$ is the least element of $M_\nu \setminus \kappa$. It suffices to show that $\kappa \notin M_\nu$. Since $M_0 \supseteq M_1 \supseteq \cdots \supseteq M_\nu \supseteq \cdots (\nu < \omega_1)$, it suffices to show that $\kappa \notin M_0$. Suppose $x \in M_0$. Then $x = t^{L_{\omega_2}}(\eta_1, \ldots, \eta_n)$ for some term $t$ and $\eta_1, \ldots, \eta_n \in \kappa \cup C_0$. Since $e \upharpoonright \kappa = id$ and $e \upharpoonright C_0 = id$, we have $e(x) = e(t^{L_{\omega_2}}(\eta_1, \ldots, \eta_n)) = t^{L_{\omega_2}}(e(\eta_1), \ldots, e(\eta_n)) = t^{L_{\omega_2}}(\eta_1, \ldots, \eta_n) = x$. Thus, $x \ne \kappa$. Hence $\kappa \notin M_0$.

(2) Let $x \in M_\tau$. Then $x = t^{L_{\omega_2}}(\eta_1, \ldots, \eta_n)$ for some term $t$ and $\eta_1, \ldots, \eta_n \in \kappa \cup C_\tau$. Since $\kappa \subseteq M_\nu$, we have $i_\nu \upharpoonright \kappa = id$. If $\eta \in C_\tau$, since $\nu < \tau$, then $\eta \in C_{\nu+1}$ and $\eta \in C_\nu$ is the $\eta$-th element of $C_\nu$ in the increasing enumeration of $C_\nu$. Thus, $i_\nu^{-1}(\eta) = \eta$ and hence $i_\nu(\eta) = \eta$. Thus $i_\nu(x) = i_\nu(t^{L_{\omega_2}}(\eta_1, \ldots, \eta_n)) = t^{L_{\omega_2}}(i_\nu(\eta_1), \ldots, i_\nu(\eta_n)) = t^{L_{\omega_2}}(\eta_1, \ldots, \eta_n) = x$.

(3) follows from (2).

(4) If $\nu < \tau$, then $M_\tau \subseteq M_\nu$ and hence $\kappa_\nu \le \kappa_\tau$. By (1), we have $\kappa_\nu > \kappa$. So applying $i_\nu$, we get $i_\nu(\kappa_\nu) > i_\nu(\kappa) = \kappa_\nu$. By (3), we have $i_\nu(\kappa_\tau) = \kappa_\tau$. Thus, $\kappa_\nu \ne \kappa_\tau$. Hence $\kappa_\nu < \kappa_\tau$. $\square$

For $\nu < \tau < \omega_1$, let $M_{\nu\tau}$ be the Skolem hull of $\kappa_\nu \cup C_\tau$ in $L_{\omega_2}$. Let $i_{\nu\tau} : L_{\omega_2} \cong M_{\nu\tau}$. Thus $i_{\nu\tau} : L_{\omega_2} \prec L_{\omega_2}$.

**Proposition 2.10** *Let $\nu < \tau < \omega_1$. Then*

(a) *if $\xi < \nu$, then $i_{\nu\tau}(\kappa_\xi) = \kappa_\xi$;*
(b) *$i_{\nu\tau}(\kappa_\nu) = \kappa_\tau$;*
(c) *if $\xi > \tau$, then $i_{\nu\tau}(\kappa_\xi) = \kappa_\xi$.*

*Proof* (a) Since $\kappa_\nu \subseteq M_{\nu\tau}$, we have $i_{\nu\tau} \upharpoonright \kappa_\nu = id$.

(b) Since $\kappa_\nu > \kappa$, we have $M_\tau \subseteq M_{\nu\tau}$ and hence $\kappa_\tau \in M_{\nu\tau}$. Since $i_{\nu\tau} \upharpoonright \kappa_\nu = id$, we have $i_{\nu\tau}(\kappa_\nu)$ is the least ordinal in $M_{\nu\tau} \geq \kappa_\nu$. Hence $\kappa_\nu \leq i_{\nu\tau}(\kappa_\nu) \leq \kappa_\tau$. It suffices to show that there is no ordinal $\delta \in M_{\nu\tau}$ such that $\kappa_\nu \leq \delta < \kappa_\tau$. Suppose that there is such a $\delta$. Then for some term $t$, we have $\delta = t^{L_{\omega_2}}(\xi_1, \ldots, \xi_n, \eta_1, \ldots, \eta_k)$ where $\xi_1, \ldots, \xi_n \in \kappa_\nu$ and $\eta_1, \ldots, \eta_k \in C_\tau$. Thus

$$L_{\omega_2} \models \exists \xi_1, \ldots, \exists \xi_n \in \kappa_\nu (\kappa_\nu \leq t(\xi_1, \ldots, \xi_n, \eta_1, \ldots, \eta_k) < \kappa_\tau).$$

Applying $i_\nu^{-1}$, since $i_\nu(\kappa_\tau) = \kappa_\tau$, $i_\nu \upharpoonright C_\tau = id$ from Proposition 2.9 (2) and $i_\nu(\kappa) = \kappa_\nu$, we have

$$L_{\omega_2} \models \exists \xi_1, \ldots, \xi_n \in \kappa (\kappa \leq t(\xi_1, \ldots, \xi_n, \eta_1, \ldots, \eta_k) < \kappa_\tau).$$

Thus, for some $\xi_1, \ldots, \xi_n \in \kappa$, we have $\kappa \leq t^{L_{\omega_2}}(\xi_1, \ldots, \xi_n, \eta_1, \ldots, \eta_k) < \kappa_\tau$. But $t^{L_{\omega_2}}(\xi_1, \ldots, \xi_n, \eta_1, \ldots, \eta_k) \in M_\tau$. This contradicts Proposition 2.9(1).

(c) If $x \in M_{\tau+1}$, then $x = t^{L_{\omega_2}}(\eta_1, \ldots, \eta_n)$ for some term $t$ and $\eta_1, \ldots, \eta_n \in \kappa \cup C_{\tau+1}$. Since $i_{\nu\tau} \upharpoonright \kappa_\nu = id$, we have $i_{\nu\tau} \upharpoonright \kappa = id$. If $\eta \in C_{\tau+1}$, then $\eta \in C_\tau$ is the $\eta$-th element of $C_\tau$ in the increasing enumeration of $C_\tau$. Thus, $i_{\nu\tau}^{-1}(\eta) = \eta$ and hence $i_{\nu\tau}(\eta) = \eta$. Thus $i_{\nu\tau}(x) = x$. In particular, $i_{\nu\tau}(\kappa_\xi) = \kappa_\xi$ for any $\xi > \tau$. □

**Proposition 2.11** *$\{\kappa_\nu \mid \nu < \omega_1\}$ is a set of indiscernibles for $L_{\omega_2}$.*

*Proof* Let $\varphi(v_1, \ldots, v_n)$ be any formula and let $\nu_1 < \cdots < \nu_n < \omega_1$, $\tau_1 < \cdots < \tau_n < \omega_1$. We show that

$$L_{\omega_2} \models \varphi(\kappa_{\nu_1}, \ldots, \kappa_{\nu_n}) \Leftrightarrow L_{\omega_2} \models \varphi(\kappa_{\tau_1}, \ldots, \kappa_{\tau_n}).$$

Take $\delta_1 < \cdots < \delta_n < \omega_1$ such that $\nu_n, \tau_n < \delta_1$. Applying $i_{\nu_n \delta_n}$, we have

$$L_{\omega_2} \models \varphi(\kappa_{\nu_1}, \ldots, \kappa_{\nu_{n-1}}, \kappa_{\nu_n}) \Leftrightarrow L_{\omega_2} \models \varphi(\kappa_{\nu_1}, \ldots, \kappa_{\nu_{n-1}}, \kappa_{\delta_n}).$$

Applying $i_{\nu_{n-1} \delta_{n-1}}$, we have

$$L_{\omega_2} \models \varphi(\kappa_{\tau_1}, \ldots, \kappa_{\nu_{n-1}}, \kappa_{\delta_n}) \Leftrightarrow L_{\omega_2} \models \varphi(\kappa_{\tau_1}, \ldots, \kappa_{\nu_{n-2}}, \kappa_{\delta_{n-1}}, \kappa_{\delta_n}).$$

Successively applying $i_{\nu_{n-2} \delta_{n-2}}, \ldots, i_{\nu_1 \delta_1}$, in the end we have

$$L_{\omega_2} \models \varphi(\kappa_{\nu_1}, \ldots, \kappa_{\nu_n}) \Leftrightarrow L_{\omega_2} \models \varphi(\kappa_{\delta_1}, \ldots, \kappa_{\delta_n}).$$

Repeating the above procedure with $\tau_1, \ldots, \tau_n$ in place of $\nu_1, \ldots, \nu_n$, we have

$$L_{\omega_2} \models \varphi(\kappa_{\tau_1}, \ldots, \kappa_{\tau_n}) \Leftrightarrow L_{\omega_2} \models \varphi(\kappa_{\delta_1}, \ldots, \kappa_{\delta_n}).$$

$\square$

**Fact 2.5**

(1) ($Z_4$, Essentially [4]) If $L_{\omega_2}$ has an uncountable set of indiscernibles, then $0^\sharp$ exists.
(2) ($Z_3$, Essentially [4]) If $L_{\omega_1}$ has an uncountable set of indiscernibles, then $0^\sharp$ exists.

**Proposition 2.12**

(1) $Z_4 + HP$ implies that there exists a nontrivial elementary embedding $j : L_{\omega_2} \prec L_{\omega_2}$ and a club $C \subseteq \omega_2$ of regular $L$-cardinals.
(2) ($Z_4$) Suppose there exists a nontrivial elementary embedding $j : L_{\omega_2} \prec L_{\omega_2}$ and a club $C \subseteq \omega_2$ of regular $L$-cardinals. Then $0^\sharp$ exists.
(3) ($Z_3$) Suppose there exists a nontrivial elementary embedding $j : L_{\omega_1} \prec L_{\omega_1}$ and a club $C \subseteq \omega_1$ of regular $L$-cardinals. Then $0^\sharp$ exists.

*Proof* Follows from the proof of Theorem 2.6 and Fact 2.5.                    $\square$

As a corollary of Theorem 5.1, $Z_3 + HP$ does not imply that $0^\sharp$ exists, and hence $Z_3 + HP$ does not imply that there exists a nontrivial elementary embedding $j : L_{\omega_1} \prec L_{\omega_1}$ and a club $C \subseteq \omega_1$ of regular $L$-cardinals.

**Corollary 2.1** ($Z_4$) *The following are equivalent:*

(1) HP.
(2) $L_{\omega_2}$ has an uncountable set of indiscernibles.
(3) $0^\sharp$ exists.

*Proof* Follows from Theorem 2.6, Fact 2.5 and the fact that $Z_2 + 0^\sharp$ exists implies HP.                                                                              $\square$

In the following, I give another proof of "HP implies that $0^\sharp$ exists" in $Z_4$ via defining $0^\sharp$ as the real which codes a countable iterable premouse.

**Theorem 2.7** (The second proof, $Z_4$) HP *implies that* $0^\sharp$ *exists.*

*Proof* Let $x$ be the witness real for HP. Now we work in $L[x]$. Take $\beta > \omega_2$ big enough such that $\beta$ is $x$-admissible and $[L_\beta[x]]^\omega \subseteq L_\beta[x]$. Take $M \prec L_\beta[x]$ such that $\omega_2 \in M$, $|M| = \omega_1$ and $M^\omega \subseteq M$, i.e. $M$ is closed under $\omega$-sequence from $M$. Let $j : L_\theta[x] \cong M \prec L_\beta[x]$ be the collapsing map. Note that $\omega_1 \leq \theta < \omega_2$, $\theta$ is $x$-admissible and $L_\theta[x]$ is closed under $\omega$-sequence. Let $\kappa = crit(j)$. Define

$U = \{X \subseteq \kappa \mid X \in L \wedge \kappa \in j(X)\}$. Since $\theta$ is an $L$-cardinal by the choice of $x$ as witnessing HP, we have $(\kappa^{+})^{L} \leq \theta < \omega_2$. Thus, $U$ is an $L$-ultrafilter on $\kappa$.

Let $\alpha = (\kappa^{+})^{L}$. Consider the structure $(L_{\alpha}, \in, U)$ which is a premouse. Since $L_{\theta}[x]$ is closed under $\omega$-sequence from $L_{\theta}[x]$, $U$ is countably complete. By Fact 2.2, we have $(L_{\alpha}, \in, U)$ is iterable. Thus, $0^{\sharp}$ exists.                              □

In a summary, $Z_4$ + HP implies that $0^{\sharp}$ exists and $Z_4$ is the minimal system in higher-order arithmetic for proving that HP and $0^{\sharp}$ exists are equivalent. From proofs of Corollary 2.1 and Theorem 2.7, the two different definitions of $0^{\sharp}$ introduced in Sect. 2.1.2 are equivalent. Ralf Schindler's book [18] also shows the equivalence of the two characterizations of $0^{\sharp}$.

Finally, I generalize Harrington's Principle to inner models $M$ and define HP($M$). I consider various known examples of inner models, and provide characterizations of HP($M$) (cf. Theorem 2.8), and also show that in some cases this generalized principle fails (cf. Corollary 2.2 and Theorem 2.9).

Recall that for limit ordinal $\alpha > \omega$, $\alpha$ is $x$-admissible if there is no $\Sigma_1(L_{\alpha}[x])$ mapping from an ordinal $\delta < \alpha$ cofinally into $\alpha$ (cf. [4, Lemma 7.2]).

**Definition 2.13**  Suppose $M$ is an inner model. *The Generalized Harrington's Principle* HP($M$) *denotes the following statement: there is a real $x$ such that, for any ordinal $\alpha$, if $\alpha$ is $x$-admissible then $\alpha$ is an $M$-cardinal (i.e. $M \models \alpha$ is a cardinal).*

Recall that Harrington's Principle HP is just HP($L$) which was isolated by Harrington in the proof of his celebrated theorem "$Det(\Sigma_1^1)$ implies that $0^{\sharp}$ exists" in [19].

**Fact 2.6**  (Theorem 21.15, [5]) *The following are equivalent:*

*(1)* $0^{\dagger}$ *exists;*
*(2)* *For every uncountable cardinal $\kappa$ there is a $\kappa$-model and a double class $\langle X, Y \rangle$ of indiscernibles for it such that: $X \subseteq \kappa$ is closed unbounded, $Y \subseteq \mathrm{Ord} \setminus (\kappa + 1)$ is a closed unbounded class, $X \cup \{\kappa\} \cup Y$ contains every uncountable cardinal and the Skolem hull of $X \cup Y$ in the $\kappa$-model is again the model.*

**Fact 2.7**  (Lemma 1.7, [20]) *Assume that $A$ is a set, $X \prec L_{\alpha}[A]$ where $\alpha \in \mathrm{Ord} \cup \{\mathrm{Ord}\}$ and the transitive closure of $A \cap L_{\alpha}[A]$ is contained in $X$. Then $X \cong L_{\alpha'}[A]$ for some $\alpha' \leq \alpha$.*

**Fact 2.8**  (Folklore) *Assume $0^{\dagger}$ exists, $L[U]$ is the unique $\kappa$-model and $\langle X, Y \rangle$ is the double class of indiscernibles for $L[U]$ as in Fact 2.6. If $\alpha \leq \kappa$ is $0^{\dagger}$-admissible, then $X$ is unbounded in $\alpha$; if $\alpha > \kappa$ is $0^{\dagger}$-admissible, then $Y$ is unbounded in $\alpha$.[5]*

**Fact 2.9**  (Exercise 21.22, [5]) *The following are equivalent:*

*(1)* $0^{\dagger}$ *exists.*

---

[5]I would like to thank W.Hugh Woodin and Sy Friedman for pointing out this fact to me. The proof of this fact is essentially similar as the proof of Theorem 1.25.

(2) *There is a $\kappa$-model for some $\kappa$ and an elementary embedding from that model to itself with critical point greater than $\kappa$.*

**Theorem 2.8** *Assume $\kappa$ is a measurable cardinal and $L[U]$ is the unique $\kappa$-model. Then* $\mathsf{HP}(L[U])$ *if and only if* $0^\dagger$ *exists.*

*Proof* ($\Rightarrow$): Let $x$ be the witness real for $\mathsf{HP}(L[U])$. Pick $\lambda > 2^\kappa$ and $X$ such that $\lambda$ is $(x, U)$-admissible, $X \prec L_\lambda[U][x]$, $|X| = 2^\kappa$, $X$ is closed under $\omega$-sequences and the transitive closure of $U \cap L_\lambda[U]$ is contained in $X$. By Fact 2.7, the transitive collapse of $X$ is of the form $L_\theta[U][x]$. Let $j : L_\theta[U][x] \cong X \prec L_\lambda[U][x]$ be the inverse of the collapsing map and $\eta = crit(j)$. Note that $\eta > \kappa$. Since $\theta$ is $(x, U)$-admissible, by $\mathsf{HP}(L[U])$, we have $\theta$ is an $L[U]$-cardinal. Define $\overline{U} = \{X \subseteq \eta \mid X \in L[U]$ and $\eta \in j(X)\}$. Since $(\eta^+)^{L[U]} \leq \theta$, we have $\overline{U} \subseteq L_\theta[U]$. $\overline{U}$ is an $L[U]$-ultrafilter on $\eta$. Since $L_\theta[U][x]$ is closed under $\omega$-sequences, we have $\overline{U}$ is countably complete. Thus, we can build a nontrivial embedding from $L[U]$ to $L[U]$ with critical point greater than $\kappa$. By Fact 2.9, $0^\dagger$ exists.

($\Leftarrow$): Suppose $0^\dagger$ exists and $\alpha$ is $0^\dagger$-admissible. We show that $\alpha$ is an $L[U]$-cardinal. By Fact 2.6, let $\langle X, Y\rangle$ be the double class of indiscernibles for $L[U]$. If $\alpha \leq \kappa$, then by Fact 2.8, we have $\alpha \in X$. If $\alpha > \kappa$, then by Fact 2.8, we have $\alpha \in Y$. Note that elements of $X$ and $Y$ are $L[U]$-cardinals.                                                    $\square$

**Fact 2.10** ([21, 22]) *Suppose there is no inner model with one measurable cardinal and let $K$ be the corresponding core model.[6] Then, $K$ has the rigidity property.*

**Corollary 2.2**

(1) *Suppose $0^\sharp$ exists. Then* $\mathsf{HP}(L[0^\sharp])$ *if and only if* $(0^\sharp)^\sharp$ *exists.*
(2) *Suppose there is no inner model with one measurable cardinal and that $K$ is the corresponding core model. Then* $\mathsf{HP}(K)$ *does not hold.*

*Proof* (1) Follows from the proof of "$\mathsf{HP} \Leftrightarrow 0^\sharp$ exists". Note that if $\alpha$ is $(0^\sharp)^\sharp$-admissible and $I$ is the class of Silver indiscernibles for $L[0^\sharp]$, then $I$ is unbounded in $\alpha$ and hence $\alpha \in I$.

(2) Note that $K = L[\mathcal{M}]$ where $\mathcal{M}$ is a class of mice. Suppose $\mathsf{HP}(K)$ holds and $x$ is the witness real for $\mathsf{HP}(K)$. Pick $\theta > \omega_2$ and $X$ such that $\theta$ is $(\mathcal{M}, x)$-admissible, $X \prec J_\theta[\mathcal{M}, x]$, $\omega_2 \in X$, $|X| = \omega_1$ and $X$ is closed under $\omega$-sequences. Since $K \models$ GCH, such an $X$ exists. By the condensation theorem for $K$, let $j : J_{\theta'}[\mathcal{M} \restriction \theta', x] \cong X \prec J_\theta[\mathcal{M}, x]$ be the inverse of the collapsing map. Let $\lambda = crit(j)$ and $U = \{X \subseteq \lambda \mid X \in K$ and $\lambda \in j(X)\}$. Note that $\theta'$ is a $K$-cardinal and $U$ is a countably complete $K$-ultrafilter on $\lambda$. Thus, there is a nontrivial elementary embedding from $K$ to $K$ which contradicts Fact 2.10.                                                    $\square$

From the proof of Corollary 2.2(2), if $M$ is an $L$-like inner model, $M \models$ CH, $M$ has the rigidity property and some proper form of condensation, then $\mathsf{HP}(M)$ does not hold.

---

[6]For the definition of $K$, I refer to [22].

**Fact 2.11** ([22], $AD^{L(\mathbb{R})}$) $HOD^{L(\mathbb{R})} = L(P)$ *for some* $P \subseteq \Theta$ *where* $\Theta = \sup\{\alpha : \exists f \in L(\mathbb{R})(f : \mathbb{R} \to \alpha$ *is surjective*$)\}$.[7]

It is an open question whether there exists a nontrivial elementary embedding from HOD to HOD.[8] However, the following fact shows that the answer to this question is negative for embeddings which are definable in $V$ from parameters.

**Fact 2.12** (Theorem 35, [23]) *Do not assume* AC. *There is no nontrivial elementary embedding from* HOD *to* HOD *that is definable in V from parameters.*[9]

**Theorem 2.9** $(ZF + AD^{L(\mathbb{R})})$ HP(HOD) *does not hold.*

*Proof* By Fact 2.11, under $ZF + AD^{L(\mathbb{R})}$, $HOD = L(P)$ for some $P \subseteq \Theta$. Suppose HP(HOD) holds. Then since $L(P) \models CH$, by a similar proof as in Corollary 2.2(2), we can show that there exists a nontrivial elementary embedding $j : L(P) \to L(P)$. Note that $j$ is definable in $V$ from parameters (i.e. there is a formula $\varphi$ and parameter $\overrightarrow{a}$ such that $j(x) = y$ if and only if $\varphi(x, y, \overrightarrow{a})$). This contradicts Fact 2.12.            $\square$

# References

1. Jech, T.J.: Set Theory. Third Millennium Edition, revised and expanded. Springer, Berlin (2003)
2. Kunen, K.: Set Theory: An Introduction to Independence Proofs. North Holland (1980)
3. Jensen, R.B., Solovay, R.M.: Some applications of almost disjoint sets. Mathematical logic and foundations of set theory. In: Proceedings of an International Colloquium Held Under the Auspices of The Israel Academy of Sciences and Humanities, vol. 59, pp. 84–104 (1970)
4. Devlin, K.J.: Constructibility. Springer, Berlin (1984)
5. Kanamori, A.: Higher Infinite: Large Cardinals in Set Theory from Their Beginnings. Springer Monographs in Mathematics, Springer, Berlin, Second edition (2003)
6. Dubose, D.A.: The equivalence of determinacy and iterated sharps. J. Symb. Log. **55**(2), 502–525 (1990)
7. Schindler, R.: Proper forcing and remarkable cardinals. Bull. Symbolic Logic **6**(2), 176–184 (2000)
8. Schindler, R.: Remarkable cardinals. Infinity, Computability, and Metamathematics (Geschke et al., eds.), Festschrift celebrating the 60th birthdays of Peter Koepke and Philip Welch, pp. 299–308
9. Magidor, M.: On the role of supercompact and extendible cardinals in logic. Israel J. Math. **10**, 147–157 (1971)
10. Schindler, R.: Proper forcing and remarkable cardinals II.: Schindler, Ralf. J. Symb. Log. **66**, 1481–1492 (2001)
11. Gitman, V., Welch, P.: Ramsey-like cardinals II. J. Symb. Log. **76**(2), 541–560 (2011)
12. Friedman, S.D.: Constructibility and Class Forcing. Chapter 8 in Handbook of Set Theory, Edited by Matthew Foreman and Akihiro Kanamori. Springer, Berlin (2010)

---

[7]The Axiom of Determinacy (AD) states that for every $A \subseteq \mathbb{R}$, the game $G_A$ is determined.

[8]The answer to this question is negative if $V = HOD$. For a very easy proof of the Kunen inconsistency in the case $V = HOD$, I refer to [23, Theorem 21].

[9]AC denotes the Axiom of Choice.

13. Cheng, Y.: Analysis of Martin-Harrington theorem in higher-order arithmetic. Ph.D. thesis, National University of Singapore (2012)
14. Cheng, Y., Schindler, R.: Harrington's Principle in higher-order arithmetic. J. Symb. Log. **80**(02), 477–489 (2015)
15. Beller, A., Jensen, R.B., Welch, P.: Coding the Universe. Cambridge University Press, Cambridge (1982)
16. Shelah, S., Stanley, L.: Coding and reshaping when there are no sharps. Set Theory of the Continuum (Judah, H. et al., Eds.), pp. 407–416. Springer, New York (1992)
17. Schindler, R.: Coding into K by reasonable forcing. Trans. Amer. Math. Soc. **353**, 479–489 (2000)
18. Schindler, R.: Set Theory: Exploring Independence and Truth. Springer (2014)
19. Harrington, L.A.: Analytic determinacy and $0^{\sharp}$. J. Symb. Log. **43**, 685–693 (1978)
20. Mitchell, W.J.: Beginning inner model theory. In: Foreman, M., Kanamori, A. (Eds.) Chapter 17 in Handbook of Set Theory. Springer, Berlin (2010)
21. Mitchell, W.J.: The covering lemma. In: Foreman, M., Kanamori, A. (Eds.) Chapter 18 in Handbook of Set Theory. Springer, Berlin (2010)
22. Steel, R.J.: An outline of inner model theory. In: Foreman, M., Kanamori, A. (Eds.) Chapter 19 in Handbook of Set Theory. Springer, Berlin (2010)
23. Hamkins, J.D., Kirmayer, G., Perlmutter, N.L.: Generalizations of the Kunen inconsistency. Ann. Pure Appl. Logic. **163**(12), 1872–1890 (2012)

# Chapter 3
# The Boldface Martin-Harrington Theorem in $Z_2$

**Abstract** The Boldface Martin-Harrington Theorem is the relativization of the Martin-Harrington Theorem. The former expresses that $Det(\Sigma_1^1)$ if and only if for any real $x$, $x^\sharp$ exists. In this chapter, I prove the Boldface Martin-Harrington Theorem in $Z_2$. In Sect. 3.1, I prove in $Z_2$ that if for any real $x$, $x^\sharp$ exists, then $Det(\Sigma_1^1)$ holds. In Sect. 3.2, I prove in $Z_2$ that $Det(\Sigma_1^1)$ implies that for any real $x$, $x^\sharp$ exists.

## 3.1 The Boldface Martin Theorem in $Z_2$

In this section, I prove the Boldface Martin Theorem in $Z_2$; recall that this theorem expresses that if $x^\sharp$ exists for any real $x$, then $Det(\Sigma_1^1)$ holds. I first prove in $Z_2$ that $0^\sharp$ exists implies $Det(\Sigma_1^1)$. By relativization, I get the Boldface Martin Theorem in $Z_2$.

**Fact 3.1** ([1, 2]) *Given $A \subseteq \omega^\omega$, then $A$ is $\Pi_1^1$ if and only if there is a recursive function $f : p \mapsto <_p$ with domain $\omega^{<\omega}$ such that*

*(1) for all $p \in \omega^{<\omega}$, $<_p$ is a linear ordering on $lh(p)$;*
*(2) for all $p \subseteq q \in \omega^{<\omega}$, $<_p$ is the restriction of $<_q$ on $lh(p)$ and*
*(3) for all $x \in \omega^\omega$, $x \in A$ iff $<_x$ is a well-ordering on $\omega$ where $<_x = \bigcup_{n \in \omega} <_{x \upharpoonright n}$.*

**Theorem 3.1** (Martin [3], $Z_2$) $0^\sharp$ *exists implies* $Det(\Sigma_1^1)$.

*Proof* We work in $Z_2$. We assume that $0^\sharp$ exists. Given a $\Pi_1^1$ set $A \subseteq \omega^\omega$, it suffices to show that the game $G_A$ is determined. By Fact 3.1, let $f : p \mapsto <_p$ be the witness function for $A$. Consider the following auxiliary game $G_A^*$:

| I | $n_0, \alpha_0$ | | $n_2, \alpha_1$ | $\cdots$ | $n_{2t}, \alpha_t$ | | $\cdots$ |
|---|---|---|---|---|---|---|---|
| II | | $n_1$ | | $n_3$ | $\cdots$ | $n_{2t+1}$ | $\cdots$ |

where player I and player II play natural numbers alternately and in addition, player I has to play countable ordinals $\alpha_0, \ldots, \alpha_m$ such that for all $m \in \omega$,

$$\langle m + 1, <_{(n_0,\dots,n_m)}\rangle \cong \langle \{\alpha_0, \dots, \alpha_m\}, \in\rangle.$$

The first player who fails to do so, loses the game; if the play is infinite then player I wins. Note that to win the game player I has to play a witness to the fact that $<_x$ is a well-ordering where $x = (n_0, n_1, \dots, n_m, \dots)$. Note that $G_A^*$ is an open game and hence determined in any inner model containing the function $f : p \mapsto <_p$. Note that since $f$ is recursive, $f \in L$.

If player I has a wining strategy $\sigma$ for $G_A^*$ in $L$, then by absoluteness, $\sigma \in L$ is also a winning strategy for player I in $V$. Thus, player I wins the game $G_A$ by following $\sigma$ avoiding the side moves $(\alpha_0, \dots, \alpha_n, \dots)$: if $x = (n_0, n_1, \dots, n_m, \dots)$ is the real by the end of a play, then $<_x$ is a well-ordering and hence $x \in A$.

Suppose player II has a winning strategy $\tau$ for $G_A^*$ in $L$. Since $0^\sharp$ exists, there exists a countable ordinal $\theta$ such that if $\alpha_0, \dots, \alpha_k$ and $\alpha_0', \dots, \alpha_k'$ are countable indiscernibles greater than $\theta$, then for any formula $\varphi(x_0, \dots, x_k)$ we have

$$L \models \varphi[\alpha_0, \dots, \alpha_k] \Leftrightarrow L \models \varphi[\alpha_0', \dots, \alpha_k'].$$

Thus, $\tau((n_0, \alpha_0), n_1, \dots, (n_{2k}, \alpha_k)) = \tau((n_0, \alpha_0'), n_1, \dots, (n_{2k}, \alpha_k'))$ for all $n_0, \dots,$ $n_{2k}$. Now we define the winning strategy $\tau'$ for player II in $G_A$. Let

$$\tau'(n_0, n_1, \dots, n_{2t}) = \tau((n_0, \alpha_0), n_1, \dots, (n_{2t}, \alpha_t)),$$

where $\alpha_0, \dots, \alpha_t$ are countable indiscernibles greater than $\theta$ with

$$\langle t + 1, <_{(n_0, n_1, \dots, n_t)}\rangle \cong \langle \{\alpha_0, \dots, \alpha_t\}, \in\rangle.$$

**Lemma 3.1** $\tau'$ is a winning strategy for player II in $G_A$.

*Proof* If not, then we get a real $x = (n_0, n_1, \dots, n_t, \dots) \in A$ from a play in which player II follows $\tau'$. Since $x \in A$ if and only if $<_x$ is a well-ordering, we can find $(\alpha_0, \dots, \alpha_n, \dots)$ of countable indiscernibles greater than $\theta$ such that

$$\langle \omega, <_x\rangle \cong \langle \{\alpha_0, \dots, \alpha_n, \dots\}, \in\rangle.$$

That is, for any $t \in \omega$, we have

$$(t + 1, <_{(n_0, n_1, \dots, n_t)}) \cong (\{\alpha_0, \dots, \alpha_t\}, \in).$$

Thus, for any $t \in \omega$, $n_{2t+1} = \tau((n_0, \alpha_0), n_1, \dots, (n_{2t}, \alpha_t))$. Since $\tau \in L$, we have $((n_0, \alpha_0), n_1, \dots, (n_{2t}, \alpha_t), \dots)$ is an infinite play in $L$ for $G_A^*$ in which player II follows $\tau$. Thus, player I wins which leads to a contradiction.                                $\square$

**Corollary 3.1** *The system* "$\mathbb{Z}_2 + 0^\sharp$ *exists*" *implies that for any* $\Sigma_1^1$ *game, either player I has a winning strategy or player II has a winning strategy recursive in* $0^\sharp$ *(and hence a* $\Delta_3^1$ *winning strategy).*

*Proof* Suppose $A \subseteq \omega^\omega$ is $\Sigma_1^1$. Associate game $G_A$ with an auxiliary game $G_A^*$ as in Theorem 3.1. From the proof of Theorem 3.1, if player II has a winning strategy $\tau$ for $G_A^*$ in $L$, then player II has a winning strategy $\tau'$ in $V$ for $G_A$ such that $\tau'$ is recursive in $0^\sharp$. □

By the relativization of "$Z_2 + 0^\sharp$ exists implies $Det(\Sigma_1^1)$", we have the Boldface Martin Theorem in $Z_2$:

**Theorem 3.2** (The Boldface Martin Theorem, $Z_2$) *If $x^\sharp$ exists for any real $x$, then $Det(\Sigma_1^1)$ holds.*

## 3.2 The Boldface Harrington Theorem in $Z_2$

In this section, I prove the Boldface Harrington Theorem in $Z_2$. Recall that this theorem expresses that if $Det(\Sigma_1^1)$, then $x^\sharp$ exists for any real $x$. I first introduce the relativized notion of HP.

**Definition 3.1** For any real $y$, let $HP(y)$ denote the statement: $\exists x \in \omega^\omega \forall \alpha$ (if $\alpha$ is a countable $x$-admissible ordinal, then $\alpha$ is an $L[y]$-cardinal).[1]

In the following, I prove in $Z_2$ that $Det(\Sigma_1^1)$ implies that for any real $x$, $x^\sharp$ exists. By the relativization of "$Z_2 + Det(\Sigma_1^1)$ implies HP", we have $Z_2 + Det(\Sigma_1^1)$ implies that $HP(x)$ holds for any real $x$. The key step in our proof is to show that $Z_2 + \forall y \in \omega^\omega(HP(y))$ implies that $y^\sharp$ exists for any real $y$.

**Theorem 3.3** $Z_2 + Det(\Sigma_1^1)$ *implies that $x^\sharp$ exists for any real $x$.*

*Proof* Relativizing the theorem "$Z_2 + Det(\Sigma_1^1)$ implies HP" to reals, $Z_2 + Det(\Sigma_1^1(x))$ implies that $HP(x)$ holds for real $x$. Thus, $Z_2 + Det(\Sigma_1^1)$ implies that for any real $x$, $HP(x)$ holds. It suffices to prove the following lemma.

**Lemma 3.2** *Assume $Z_2 + \forall y \in \omega^\omega(HP(y))$. Then for any real $y$, $y^\sharp$ exists.*

*Proof* Fix a real $y$. Let $z$ be a witness real for $HP(y)$ with $y \leq_T z$. Let $s$ be a witness real for $HP(z)$ with $z \leq_T s$. Since $Z_2 \vdash \forall \alpha \forall x \in \omega^\omega \exists \beta(\beta$ is the $\alpha$-th countable $x$-admissible ordinal), there exists an increasing sequence $\langle \lambda_i \mid i \in \omega \rangle$ of countable $s$-admissible ordinals. Thus, for any $i \in \omega$, $\lambda_i$ is an $L[z]$-cardinal. Let $\lambda = \omega_3^{L[z]}$. From the definition of $\lambda$ and $Z_4$, we have $L_\lambda[z] \models Z_4$. Since $z$ witnesses $HP(y)$ and $y \leq_T z$, we have $L_\lambda[z] \models HP(y)$. By relativizing the theorem "$Z_4 + HP$ implies that $0^\sharp$ exists" to real $y$, we have $Z_4 + HP(y)$ implies that $y^\sharp$ exists. Thus, $L_\lambda[z] \models y^\sharp$ exists.

---

[1]Note that for fixed real $y$, $Z_2 + HP(y)$ implies that there is no largest $L[y]$-cardinal and hence $L[y]$ is a model of ZFC.

**Fact 3.2** ([1, 3])

*(1)* $Z_2 \vdash \forall x \in \omega^\omega \, \forall y \in \omega^\omega (if \; \omega_1^{L[x]} \; exists \; and \; L_\lambda[x] \models$ "$y^\sharp$ *exists*" *for some* $\lambda \geq$
$\omega_1^{L[x]}$, *then* $L[x] \models$ "$y^\sharp$ *exists*").
*(2)* $Z_2 \vdash \forall x \in \omega^\omega \, \forall y \in \omega^\omega (if \; L[x] \models y^\sharp$ *exists, then* $y^\sharp$ *exists*).

Note that "$y^\sharp$ exists" is $\Sigma_3^1(y)$. Since $\omega_1^{L[z]} < \lambda$, by Fact 3.2(1), $L[z] \models y^\sharp$ exists.
By Fact 3.2(2), $y^\sharp$ exists.                                                   $\square$

From Theorems 3.2 and 3.3, the Boldface Martin-Harrington Theorem is provable
in $Z_2$.

**Theorem 3.4** (The Boldface Martin-Harrington Theorem, $Z_2$) $Det(\Sigma_1^1)$ *if and only
if for any real* $x$, $x^\sharp$ *exists.*

Finally, I give a summary of the main results we have obtained about the analysis
of the Martin-Harrington Theorem in higher-order arithmetic.

- The Boldface Martin-Harrington Theorem is provable in $Z_2$.
- The Martin's Theorem is provable in $Z_2$.
- The theorem "$Det(\Sigma_1^1)$ implies HP" is provable in $Z_2$.
- The theorem "HP implies that $0^\sharp$ exists" is neither provable in $Z_2$ nor provable in
  $Z_3$, but it is provable in $Z_4$.

As far as I know, the question *Is Harrington's Theorem provable in* $Z_2$? is still open
and Woodin conjectures that the Harrington's Theorem is provable in $Z_2$.

# References

1. Kanamori, A.: Higher Infinite: Large Cardinals in Set Theory from Their Beginnings. 2nd Edn.
   Springer Monographs in Mathematics, Springer, Berlin (2003)
2. Schindler, R.: Set Theory: Exploring Independence and Truth. Springer (2014)
3. Jech, T.J.: Set Theory. Third Millennium Edition, revised and expanded. Springer, Berlin (2003)

# Chapter 4
# Strengthenings of Harrington's Principle

**Abstract** In this chapter, we examine the large cardinal strength of strengthenings of Harrington's Principle, $HP(\varphi)$, over $Z_2$ and $Z_3$. In Sect. 4.3, we prove that $Z_2 + HP(\varphi)$ is equiconsistent with "$ZFC + \{\alpha | \varphi(\alpha)\}$ is stationary". In Sect. 4.5, we prove that $Z_3 + HP(\varphi)$ is equiconsistent with "$ZFC+$ there exists a remarkable cardinal $\kappa$ with $\varphi(\kappa) + \{\alpha | \varphi(\alpha) \wedge \{\beta < \alpha | \varphi(\beta)\}$ is stationary in $\alpha\}$ is stationary". As a corollary, $Z_4$ is the minimal system of higher-order arithmetic for proving that $HP$, $HP(\varphi)$, and $0^\sharp$ exists are pairwise equivalent with each other.

## 4.1 Overview

In this Chapter, we extend the work on the large cardinal strength of $HP$ over $Z_2$ and $Z_3$ in Chap. 2 to the large cardinal strength of the strengthenings of $HP$ over $Z_2$ and $Z_3$. The structure of this chapter is as follows. In Sect. 4.2, I introduce Harrington's club shooting forcing used in Sect. 4.3. In Sect. 4.3, we prove the large cardinal strength of $Z_2 + HP(\varphi)$. In Sect. 4.4, I introduce subcomplete forcing used in Sect. 4.5. In Sect. 4.5, we establish the large cardinal strength of $Z_3 + HP(\varphi)$. Sections 4.3 and 4.5 in this chapter are based on material from [1] with considerable revisions and improvements.

## 4.2 Harrington's Club Shooting Forcing

In this section, I introduce Harrington's club shooting forcing $\mathbb{P}_S$ (cf. [2]) where $S$ is a stationary subset of $\omega_1$. If $V[G]$ is a generic extension, then every club $C \in V$ on $\omega_1$ remains a club in $V[G]$, provided that $\omega_1$ is preserved. But a stationary set $S$ in $V$ may no longer be stationary in $V[G]$ since there may be a club $C \in V[G]$ disjoint from $S$. In this section, we show that forcing with $\mathbb{P}_S$ shoots a club $C$ such that $C \subseteq S$ and adds no new reals.

Y. Cheng, *Incompleteness for Higher-Order Arithmetic*, SpringerBriefs
in Mathematics, https://doi.org/10.1007/978-981-13-9949-7_4

**Definition 4.1** ([2]) $\mathbb{P}_S = \{p : p$ is a closed bounded subset of $\omega_1$ and $p \subseteq S\}$. For $p, q \in \mathbb{P}_S$, $p \leq q$ if $p \supseteq q$ and $sup(q) < \alpha$ for any $\alpha \in p \setminus q$.

**Lemma 4.1** *If $G$ is $\mathbb{P}_S$-generic over $V$, then $V[G] \models \bigcup G$ is a club on $\omega_1$.*

*Proof* We work in $V[G]$. We first show that $\bigcup G$ is unbounded in $\omega_1$. Fix $\alpha < \omega_1$. $D_\alpha = \{p \in \mathbb{P}_S \mid sup(p) > \alpha\}$ is dense in $\mathbb{P}_S$. Thus, there is $p \in G$ such that $sup(p) > \alpha$. Since $p$ is closed, $sup(p) \in p$ and hence $sup(p) \in \bigcup G$. Then we show that $\bigcup G$ is closed. Assume $\alpha$ is a limit point of $\bigcup G$. Since $\alpha < \omega_1$, there is $\beta \in \bigcup G$ such that $\beta > \alpha$. Let $\beta \in q \in G$. Thus, there is $q \in G$ such that $\alpha \in \bigcup q = sup(q)$. I show that $\alpha \cap \bigcup G \subseteq q$. Let $\gamma \in \alpha \cap \bigcup G$ and $\gamma \in q' \in G$. If $q$ end extends $q'$, then $\gamma \in q$. If $q'$ end extends $q$, then $\gamma \in q$ also holds (if $\gamma \notin q$, then $\gamma > sup(q)$ since $\gamma \in q' \setminus q$. But $\gamma < \alpha < sup(q)$ which leads to a contradiction). Since $\alpha$ is a limit point of $\bigcup G$, $\alpha \cap \bigcup G \subseteq q$ and $q$ is closed, $\alpha \in q$ and thus $\alpha \in \bigcup G$.  □

Now we show that $\omega_1$ is preserved and hence $\omega_1^{V[G]} = \omega_1^V$.

**Lemma 4.2** *$\mathbb{P}_S$ is $\omega$-distributive and hence preserves $\omega_1$.*

*Proof* It suffices to show that any countable set of ordinals in $V[G]$ is in the ground model. Let $p \Vdash \dot{f} : \omega \to Ord$. It suffices to show that there exist $q \leq p$ and $g$ such that $q \Vdash \dot{f} = \check{g}$.

By induction on $\alpha$, we construct a chain $\{A_\alpha \mid \alpha < \omega_1\}$ of countable subsets of $\mathbb{P}_S$. Let $A_0 = \{p\}$. If $\alpha$ is a limit ordinal, $A_\alpha = \bigcup_{\beta < \alpha} A_\beta$. Given $A_\alpha$, let $\gamma_\alpha = sup(\{sup(q) : q \in A_\alpha\})$. Since $A_\alpha \subseteq \mathbb{P}_S$ is countable, we have $\gamma_\alpha < \omega_1$. For each $q \in A_\alpha$ and each $n \in \omega$, choose some $r = r(q, n) \in \mathbb{P}_S$ such that $r \leq q$, $r$ decides $\dot{f}(n)$ and $sup(r) > \gamma_\alpha$.[1] Let $A_{\alpha+1} = A_\alpha \cup \{r(q, n) : q \in A_\alpha, n \in \omega\}$. The sequence $\{\gamma_\alpha : \alpha < \omega_1\}$ is increasing and continuous. Let

$$C = \{\lambda < \omega_1 : \alpha < \lambda \to \gamma_\alpha < \lambda\}.$$

Since $C$ is a club on $\omega_1$ and $S$ is stationary, there exists a limit ordinal $\lambda \in C \cap S$. Let $\{\alpha_n : n \in \omega\}$ be an increasing sequence with limit $\lambda$. By the definition of $C$, $\lim_n \gamma_{\alpha_n} = \lambda$. From the construction of $\{A_\alpha : \alpha < \omega_1\}$, there is a sequence $\{p_n : n \in \omega\}$ such that $p_0 = p$ and for any $n \in \omega$, $p_{n+1} \in A_{\alpha_n+1}$, $p_{n+1} \leq p_n$, $\gamma_{\alpha_n} < sup(p_{n+1})$ and $p_{n+1}$ decides $\dot{f}(n)$. Since $sup(p_{n+1}) \leq \gamma_{\alpha_n+1}$, $\lim_{n \in \omega} sup(p_n) = \lim_n \gamma_{\alpha_n} = \lambda$. Since $\lambda \in S$, $q = \bigcup_{n \in \omega} p_n \cup \{\lambda\}$ is closed and $q \subseteq S$. Thus, $q \in \mathbb{P}_S$. Since $q \leq p_n$ for all $n \in \omega$, $q$ decides each $\dot{f}(n)$. For $n \in \omega$, let $g(n)$ be the value of $\dot{f}(n)$ decided by $q$. Thus, $q \Vdash \dot{f} = \check{g}$.  □

**Theorem 4.1** (Baumgartner, Harrington and Kleinberg, [2]) *There exists an $\omega_1$-preserving generic extension $V[G]$ such that $V[G] \models \exists C \subseteq S (C$ is a club on $\omega_1$).*

*Proof* Follows from Lemmas 4.1 and 4.2.  □

---

[1] $r$ decides $\dot{f}(n)$ means for some $x \in V, r \Vdash \dot{f}(n) = \check{x}$.

*Remark 4.1* For stationary $S \subseteq \omega_1$, if $\omega_1 \setminus S$ is also stationary, then the stationarity of $\omega_1 \setminus S$ is destroyed by $\mathbb{P}_S$ since $\omega_1 \setminus S$ is disjoint from the new added club $C \subseteq S$ in $V[G]$.

By way of a summary, the class $\mathbb{P}_S$ has the following properties:

(1) $\mathbb{P}_S$ is not proper;
(2) $\mathbb{P}_S$ is not $\omega_1$-closed;
(3) $\mathbb{P}_S$ is $\omega$-distributive and adds no new reals;
(4) $|\mathbb{P}_S| = 2^\omega$. Assuming **CH**, $\mathbb{P}_S$ preserves all cardinals;[2]
(5) Suppose $G$ is $\mathbb{P}_S$-generic over $V$ and $C \subseteq S$ is the new club in $V[G]$. Then $C \cap \alpha \in V$ for all $\alpha < \omega_1$.

## 4.3 The Strength of $Z_2 + HP(\varphi)$

In this section, we examine the large cardinal strength of $HP(\varphi)$, the strengthing of $HP$, over $Z_2$. We prove that $Z_2 + HP(\varphi)$ is equiconsistent with $ZFC + Ord$ is $\varphi$-Mahlo. Let us first introduce the two key notions: $HP(\varphi)$ and 'Ord is $\varphi$-Mahlo'.

**Definition 4.2** Let $\varphi(-)$ be a $\Sigma_2$-formula in $\mathfrak{L}_{st}$ such that, provably in $ZFC$: for all $\alpha$, if $\varphi(\alpha)$, then $\alpha$ is an inaccessible cardinal and $L \models \varphi(\alpha)$. Let $HP(\varphi)$ denote the statement: $\exists x \in \omega^\omega \forall \alpha$ (if $\alpha$ is a $x$-admissible ordinal, then $L \models \varphi(\alpha)$).

Let us give some examples of such $\varphi(-)$: $\kappa$ is inaccessible, Mahlo, weakly compact, $\Pi_m^n$-indescribable, totally indescribable, $n$-subtle, $n$-ineffable, totally ineffable cardinal, $\alpha$-iterable $(\alpha < \omega_1^L)$ and $\alpha$-Erdös cardinal $(\alpha < \omega_1^L)$. However, $\kappa$ being reflecting, unfoldable or remarkable cannot be expressed in a $\Sigma_2$ fashion.[3]

**Definition 4.3** Let $\varphi(-)$ be as in Definition 4.2. Let $\delta$ be an inaccessible cardinal or $\delta = Ord$. We say that $\delta$ is $\varphi$-Mahlo if $\{\alpha < \delta \mid \varphi(\alpha)\}$ is stationary in $\delta$.

Note that we do not require a $\varphi$-Mahlo to satisfy $\varphi(-)$. In this section, we assume that $\varphi(-)$ is the formula satisfying the hypothesis as stated in Definition 4.2.

**Theorem 4.2** *Let $\varphi(-)$ be as in Definition 4.2. The following two theories are equiconsistent:*

*(1) $Z_2 + HP(\varphi)$.*
*(2) $ZFC + Ord$ is $\varphi$-Mahlo.*

---

[2]If $2^\omega > \omega_1$, then $2^\omega$ is collapsed to $\omega_1$ in $V[G]$ where $G$ is $\mathbb{P}_S$-generic over $V$.
[3]For remarkable cardinal, I refer to Sect. 2.1.3. For definitions of other notions of large cardinals, I refer to Appendix C.

*Proof* Let us first assume (1), and let $x \in \omega^\omega$ be as in $\mathsf{HP}(\varphi)$. There is a club class of $x$-admissibles, so that $\{\alpha : L \models \varphi(\alpha)\}$ contains a club. Hence $L \models$ "ZFC $+ \{\alpha \in$ Ord $: \varphi(\alpha)\}$ is stationary". This shows that (2) holds in $L$.

Let us now assume (2). We force over $L$. Let $S = \{\alpha \in$ Ord $: \varphi(\alpha)\}$. Let $G$ be $Col(\omega, <$ Ord$)$-generic over $L$. Then $L[G] \models \mathsf{Z}_2$, and in $L[G]$, $S$ is still stationary since $Col(\omega, <$ Ord$)$ has the Ord-c.c. We can thus shoot a club through $S$ via $\mathbb{P} = \{p : p$ is a closed set of ordinals and $p \subseteq S\}$. Let $H$ be $\mathbb{P}$-generic over $L[G]$. Standard arguments give that $\mathbb{P}$ is $\omega$-distributive, which implies that $L[G][H] \models \mathsf{Z}_2$. Let $C \subseteq S$ be the club added by $H$. We may pick $A \subseteq$ Ord such that $L[G][H] = L[A]$.

We now reshape $A$ as follows.[4] Let $p \in \mathbb{T}$ iff $p : \alpha \to 2$ for some ordinal $\alpha$ such that for all $\xi \leq \alpha$,

$$L_{\xi+1}[A \cap \xi, p \restriction \xi] \models \xi \text{ is countable}.$$

We claim that $\mathbb{T}$ is $\omega$-distributive. To see this, let $(D_n | n < \omega)$ be a, say, $\Sigma_m$-definable sequence of open dense classes, and let $p \in \mathbb{T}$. Let $E$ be the class of all $\beta$ such that $L_\beta[G][H] \prec_{\Sigma_{m+5}} L[G][H]$ and $p$ as well-as the parameters defining $(D_n | n < \omega)$ are all in $L_\beta[G][H]$. $E$ is club, and we may let $\alpha$ be the $\omega$-th element of $E$. Then $E \cap \alpha$ is $\Sigma_{m+6}$-definable over $L_\alpha[G][H]$ and cofinal in $\alpha$, so that $\alpha$ has cofinality $\omega$ in $L_{\alpha+1}[G][H]$. A much simplified variant of the argument from Lemma 2.3, which we will leave as an exercise to the reader, then produces some $q \in \mathbb{T}$ with $q \leq p$, $q : \alpha \to 2$ and $q \in \bigcap_{n \in \omega} D_n$.

Let $K$ be $\mathbb{T}$-generic over $L[G][H]$. In $L[G][H][K]$, we may then pick some $B \subseteq$ Ord such that $L[G][H][K] = L[B]$, if $\lambda \in C \setminus (\omega + 1)$, then $B \cap [\lambda, \lambda + \omega)$, restricted to the odd ordinals, codes a well-ordering of $min(C \setminus (\lambda + 1))$, and for all $\alpha \geq \omega$, we have

$$L_{\alpha+1}[B \cap \alpha] \models \alpha \text{ is countable}. \tag{4.1}$$

We may now continue as in the proof of Theorem 2.8.

We do standard almost disjoint forcing to add a real $x$ such that if $(c_\alpha : \alpha \in$ Ord$)$ is the canonical sequence of pairwise almost disjoint subsets of $\omega$ given by (4.1), then for any $\alpha \in$ Ord, we have $\alpha \in B \Leftrightarrow |x \cap c_\alpha| < \omega$. In particular, $L[B][x] = L[x]$. This forcing is c.c.c., so that $L[x] \models \mathsf{Z}_2$.

We claim that in $L[x]$, we have $\mathsf{HP}(\varphi)$ holds. It suffices to show that if $\alpha$ is $x$-admissible, then $\alpha \in C$. Suppose $\alpha$ is $x$-admissible but $\alpha \notin C$. Let $\lambda$ be the largest element of $C$ such that $\lambda < \alpha$. Note that we can define $B \cap \alpha$ over $L_\alpha[x]$. Since $B \cap [\lambda, \lambda + \omega) \in L_\alpha[x]$ and $B \cap [\lambda, \lambda + \omega)$, restricted to the odd ordinals, codes a well-ordering of $min(C \setminus (\lambda + 1))$, we have $min(C \setminus (\lambda + 1)) \in L_\alpha[x]$ since $\alpha$ is $x$-admissible. But $min(C \setminus (\lambda + 1)) > \alpha$, which leads to a contradiction. Thus, $L[x] \models \mathsf{Z}_2 + \mathsf{HP}(\varphi)$. $\qquad\square$

**Corollary 4.1** $\mathsf{Z}_2 + \mathsf{HP}(\varphi)$ *does not imply that* $0^\sharp$ *exists.*

---

[4]In the proof of Theorem 2.8 there was no need for reshaping due to (2.3).

## 4.4 Subcomplete Forcing

In this section, I introduce Jensen's theory of *subcomplete forcing* and *Revised Countable Support* (RCS) iteration, to be used in Sect. 4.5. For Jensen's theory of subcomplete forcing, I refer to [3]. For Revised Countable Support (RCS) iteration, I refer to [4] and [3]. All definitions and facts in this section are due to Jensen (cf. [3] and [4]).

**Definition 4.4** ([3])

(1) Let $N$ be transitive. $N$ is full if $\omega \in N$ and there is $\gamma$ such that $L_\gamma(N) \models \mathsf{ZFC}^-$ and $N$ is regular in $L_\gamma(N)$, that is if $f : x \to N, x \in N$ and $f \in L_\gamma(N)$, then $ran(f) \in N$.
(2) Let $\mathbb{B}$ be a complete Boolean algebra. Let $\delta(\mathbb{B})$ be the smallest cardinality of a set which lies dense in $\mathbb{B} \setminus \{0\}$.
(3) Let $N = L_\gamma^A = \langle L_\gamma[A], \in, A \cap L_\gamma[A] \rangle$ be a model of $\mathsf{ZFC}^-$. Let $X \cup \{\delta\} \subseteq N$. Define $C_\delta^N(X) =$ the smallest $Y \prec N$ such that $X \cup \{\delta\} \subseteq Y$.

**Definition 4.5** ([3]) Let $\mathbb{B}$ be a complete Boolean algebra. $\mathbb{B}$ is a subcomplete forcing if for sufficiently large cardinals $\theta$ we have: $\mathbb{B} \in H_\theta$ and for any $\mathsf{ZFC}^-$ model $N = L_\tau^A$ such that $\theta < \tau$ and $H_\theta \subseteq N$ we have: Let $\sigma : \overline{N} \prec N$ where $\overline{N}$ is countable and full. Let $\sigma(\overline{\theta}, \overline{s}, \overline{\mathbb{B}}) = \theta, s, \mathbb{B}$ where $\overline{s} \in \overline{N}$. Let $\overline{G}$ be $\overline{\mathbb{B}}$-generic over $\overline{N}$. Then there is $b \in \mathbb{B} \setminus \{0\}$ such that whenever $G$ is $\mathbb{B}$-generic over $V$ and $b \in G$, there is $\sigma' \in V[G]$ such that

(a) $\sigma' : \overline{N} \prec N$,
(b) $\sigma'(\overline{\theta}, \overline{s}, \overline{\mathbb{B}}) = \theta, s, \mathbb{B}$,
(c) $C_\delta^N(ran(\sigma')) = C_\delta^N(ran(\sigma))$ where $\delta = \delta(\mathbb{B})$,
(d) $\sigma'{}''\overline{G} \subseteq G$.

In the following, we give some examples of subcomplete forcing notions we will use in Sect. 4.5. The set $\omega_2^{<\omega}$ of monotone finite sequences in $\omega_2$ is a tree ordered by inclusion. Namba forcing is the collection of all subtrees $T \neq \emptyset$ of $\omega_2^{<\omega}$ with a unique stem, stem($T$), such that every element of $T$ is comparable with stem($T$) and every element extending stem($T$) has $\omega_2$ immediate successors in $T$. The order is defined by: $T \leq \overline{T}$ if $T \subseteq \overline{T}$. If $G$ is generic over Namba forcing, then $S = \bigcup \bigcap G$ is a cofinal map of $\omega$ into $\omega_2^V$. We call any such $S$ a Namba sequence. Namba forcing is stationary set preserving and adds no reals if $\mathsf{CH}$ holds.

**Lemma 4.3** (Lemma 6.2, [3]) *Assume* $\mathsf{CH}$. *Then Namba forcing is subcomplete.*

**Definition 4.6** ([3]) Suppose $\kappa$ is a cardinal or $\kappa = \mathrm{Ord}$. Define $Club(\kappa, S) = \{p : \alpha + 1 \to S$ for some $\alpha < \kappa$ and $p$ is increasing and continuous$\}$. The extension relation is defined by: $p \leq q$ if and only if $p \supseteq q$.

The forcing notion $Club(\omega_1, S)$ will be used in the proof of Theorem 4.5. If $G$ is $Club(\omega_1, S)$-generic, then $\bigcup G : \omega_1 \to S$ is increasing, continuous and cofinal in $S$.

**Lemma 4.4** (Lemma 6.3, [3]) *Let $\kappa > \omega_1$ be a regular cardinal. Let $S \subseteq \kappa$ be a stationary set. Then $Club(\omega_1, S)$ is subcomplete.*

**Lemma 4.5** (Lemma 18.6, [5]) *Suppose* **CH** *holds and $S \subseteq \omega_2$ is such that $\{\alpha \in S \cap cf(\omega_1)$: there exists $C \subseteq S \cap \alpha$ such that $C$ is a club in $\alpha\}$ is stationary. Then $Club(\omega_2, S)$ is $\omega_1$-distributive.*

**Definition 4.7** By an iteration of length $\alpha > 0$ we mean a sequence $\langle \mathbb{B}_i : i < \alpha \rangle$ of complete BA's such that (1) $\mathbb{B}_i \subseteq \mathbb{B}_j$ for $i \leq j < \alpha$; (2) if $\lambda < \alpha$ is a limit ordinal, then $\mathbb{B}_\lambda$ is generated by $\bigcup_{i<\lambda} \mathbb{B}_i$, i.e. there is no proper $B \subset \mathbb{B}_\lambda$ s.t. $\bigcup_{i<\lambda} \mathbb{B}_i \subset B$ and $\bigcap X, \bigcup X \in B$ for all $X \subset B$.

If $G_i$ is $\mathbb{B}_{i+1}$-generic and $G_i = G \cap \mathbb{B}_i$, then $V[G] = V[G_i][\tilde{G}_i]$ where $\tilde{G}_i = \{b/G_i | b \in G\}$ is $\tilde{\mathbb{B}}_i = \mathbb{B}_{i+1}/G_i$-generic. If $G$ is $\lambda$-generic for a limit $\lambda$, then $V[G]$ can be regarded as a limit of successive $\tilde{\mathbb{B}}_i$-generic extensions, where $G_i = G \cap \mathbb{B}_i$, $\tilde{\mathbb{B}}_i = \mathbb{B}_{i+1}/G_i$ for $i < \lambda$. In practice, we usually at the i-th stage pick a $\dot{\mathbb{B}}_i$ s.t. $\Vdash_{\mathbb{B}_i} (\dot{\mathbb{B}}_i$ is a complete **BA**), and arrange that: $\Vdash_{\mathbb{B}_i} (\check{\mathbb{B}}_i/\dot{G}$ is isomorphic to $\dot{\mathbb{B}})$. If the construction of the $\mathbb{B}_i$'s is sufficiently canonical, then the iteration is completely characterized by the sequence of $\dot{\mathbb{B}}_i$'s. However, our definition of iteration gives us great leeway in choosing $\mathbb{B}_\lambda$ for limit $\lambda < \alpha$.

By a thread in $\langle \mathbb{B}_i : i < \lambda \rangle$ we mean a $b = \langle b_i : i < \lambda \rangle$ s.t. $b_j \in \mathbb{B}_j \setminus \{0\}$ and $\bigcap \{b \in \mathbb{B}_i | b \supseteq b_j\} = b_i$ for $i \leq j < \lambda$. We call $\mathbb{B}_\lambda$ an inverse limit of $\langle \mathbb{B}_i : i < \lambda \rangle$ if the following hold:

(1) if $b$ is a thread, then $b^* = \bigcap_{i<\lambda} b_i \neq 0$ in $\mathbb{B}_\lambda$; and
(2) the set of such $b^*$ is dense in $\mathbb{B}_\lambda$.

By an **RCS** thread we mean a thread $b$ such that either $b_i \Vdash_{\mathbb{B}_i}$ "$cf(\check{\lambda}) = \omega$" for some $i < \lambda$ or the support of $b$ is bounded in $\lambda$. The **RCS** limit is then defined like the inverse limit, using only **RCS** threads. An **RCS** iteration is one which uses the **RCS** limit at all limit points.

**Theorem 4.3** (Theorem 2, [3]) *Let $\langle \mathbb{B}_i : i < \omega \rangle$ be such that $\mathbb{B}_0 = 2$, $\mathbb{B}_i \subseteq \mathbb{B}_{i+1}$ and $\Vdash_i (\check{\mathbb{B}}_{i+1}/\dot{G}$ is subcomplete) for $i < \omega$. Let $\mathbb{B}_\omega$ be the inverse limit of $\langle \mathbb{B}_i : i < \omega \rangle$. Then $\mathbb{B}_\omega$ is subcomplete.*

Theorem 4.3 can be generalized to countable support iterations of length $< \omega_2$. At $\omega_2$ it can fail, however, since in a countable support iteration we are required to take a direct limit at $\omega_2$. If some earlier stage changed the cofinality of $\omega_2$ to $\omega$ (e.g. if $\mathbb{B}_1$ were Namba forcing), then the direct limit would not be subcomplete. Hence for longer iterations we must employ revised countable support iterations.

**Theorem 4.4** (Theorem 3, [3]) *Let $\langle \mathbb{B}_i : i < \alpha \rangle$ be an **RCS**-iteration such that for all $i + 1 < \alpha$:*

(a) $\mathbb{B}_i \neq \mathbb{B}_{i+1}$;
(b) $\Vdash_i (\check{\mathbb{B}}_{i+1}/\dot{G}$ is subcomplete); and
(c) $\Vdash_{i+1} (\delta(\check{\mathbb{B}}_i)$ has cardinality $\leq \omega_1)$.

*Then every $\mathbb{B}_i$ is subcomplete.*

By [3] (cf. also [6]), subcomplete forcings add no reals. RCS iterations are particulary suited to subcomplete forcing. From Theorem 4.4, subcomplete forcings are closed under Revised Countable Support (RCS) iterations subject to the usual constraints. We will use this fact in our proof of Theorem 4.5.

## 4.5 The Strength of $Z_3 + HP(\varphi)$

In this section, we examine the large cardinal strength of $HP(\varphi)$, the strengthing of $HP$, over $Z_3$. We prove that $Z_3 + HP(\varphi)$ is equiconsistent with "ZFC+ there exists a remarkable cardinal $\kappa$ with $\varphi(\kappa) + Ord$ is 2-$\varphi$-Mahlo". Let us first introduce the key notion, namely 2-$\varphi$-Mahlo.

**Definition 4.8** Let $\varphi(-)$ be as in Definition 4.2. Let $\delta$ be an inaccessible cardinal or $\delta = Ord$. We say that $\delta$ is 2-$\varphi$-Mahlo iff $\{\alpha < \delta | \varphi(\alpha) \wedge \{\beta < \alpha | \varphi(\beta)\}$ is stationary in $\alpha\}$ is stationary in $\delta$.

Notice that we do not require a 2-$\varphi$-Mahlo to satisfy $\varphi(-)$. In this section, we assume that $\varphi(-)$ is the formula satisfying the hypothesis as stated in Definition 4.2.

**Proposition 4.1** $Z_3 + HP(\varphi)$ *implies that* $L \models$ "ZFC+ *there exists a remarkable cardinal* $\kappa$ *with* $\varphi(\kappa) + Ord$ *is 2-$\varphi$-Mahlo".*

*Proof* Since $HP(\varphi)$ implies $HP$, from Proposition 2.7, we have $Z_3 + HP(\varphi)$ implies $L \models$ "ZFC $+ \omega_1^V$ is remarkable". Let $x \in \omega^\omega$ witness $HP(\varphi)$. As $\omega_1^V$ is $x$-admissible, we have $\varphi(\omega_1^V)$ holds true in $L$. There is a club of $x$-admissibles, so that we may pick some club $C \subseteq \{\alpha \in Ord : L \models \varphi(\alpha)\}$. Suppose $D$ is a club in $L$. Pick $\alpha$ in $C \cap D$ of cofinality $\omega_1$ such that $\alpha$ is a limit point of $C \cap D$. Since $\alpha \in C$, we have $L \models \varphi(\alpha)$. We want to show that $\{\beta < \alpha : L \models \varphi(\beta)\}$ is stationary in $L$. Let $E \subseteq \alpha$ in $L$ be a club in $\alpha$. Note that $E \cap C \cap \alpha \neq \emptyset$. If $\beta \in E \cap C \cap \alpha$, then $L \models \varphi(\beta)$. Hence $Ord$ is 2-$\varphi$-Mahlo in $L$. $\square$

**Theorem 4.5** *If "ZFC+ there exists a remarkable cardinal* $\kappa$ *with* $\varphi(\kappa) + Ord$ *is 2-$\varphi$-Mahlo" is consistent, then "$Z_3 + HP(\varphi)$" is consistent.*

*Proof* We force over $L$. Suppose that "ZFC+ there exists a remarkable cardinal $\kappa$ with $\varphi(\kappa) + Ord$ is 2-$\varphi$-Mahlo" holds in $L$. Let $H$ be $Col(\omega, < \kappa)$-generic over $L$.

**Lemma 4.6** $\{\alpha < \kappa : L \models \varphi(\alpha)\}$ *is stationary in* $L[H]$.

*Proof* We work in $L[H]$. Let $C \subseteq \kappa = \omega_1^{L[H]}$ be club, and let $L_\theta \models \varphi(\kappa)$, where $\theta > \kappa$ is regular. As $\kappa$ is remarkable, there is some $\sigma : L_{\bar{\theta}}[H \cap L_\alpha] \to L_\theta[H]$ such that $\alpha = crit(\sigma), \sigma(\alpha) = \kappa, C \in ran(\sigma)$ and $\bar{\theta}$ is a regular cardinal in $L$. By elementarity, we have $L_{\bar{\theta}} \models \varphi(\alpha)$, which implies that $L \models \varphi(\alpha)$ since $\varphi$ is $\Sigma_2$. But $\alpha \in C$. $\square$

Let $H$ be $Col(\omega, < \kappa)$-generic over $L$. Over $L[H]$, we define a class RCS-iteration $((\mathbb{P}_\alpha, \dot{\mathbb{Q}}_\alpha) : \alpha \in \mathsf{Ord})$ as follows. We let $\mathbb{P}_0 = \emptyset$, $\mathbb{P}_{\alpha+1} = \mathbb{P}_\alpha * \dot{\mathbb{Q}}_\alpha$ for $\alpha \in \mathsf{Ord}$ and for limit ordinal $\alpha$ we let $\mathbb{P}_\alpha$ be the revised limit $\langle \langle \mathbb{P}_\beta, \dot{\mathbb{Q}}_\beta \rangle : \beta \in \alpha \rangle$. The definition of $\mathbb{Q}_\alpha$ splits into three cases as follows. Let

(1)  $S_0 = \{\alpha : L \models \neg\varphi(\alpha)\}$,
(2)  $S_1 = \{\alpha : L \models \varphi(\alpha)$, but $\{\beta < \alpha : \varphi(\beta)\}$ is not stationary in $L\}$, and
(3)  $S_2 = \{\alpha : L \models \varphi(\alpha)$, but $\{\beta < \alpha : \varphi(\beta)\}$ is stationary in $L\}$.

Case 1  If $\alpha \in S_0$, then let $\mathbb{Q}_\alpha = Col(\omega_1, 2^{\omega_1})$ which collapses $2^{\omega_1}$ to $\omega_1$ by countable conditions.
Case 2  If $\alpha \in S_1$, then let $\mathbb{Q}_\alpha =$ Namba forcing.
Case 3  If $\alpha \in S_2$, then let $\mathbb{Q}_\alpha = Club(\omega_1, S_1 \cap \alpha)$.

Note that if $L \models \varphi(\alpha)$, then $L^{Col(\omega, <\kappa)*\mathbb{P}_\alpha} \models \alpha = \omega_2$ since $Col(\omega, < \kappa) * \mathbb{P}_\alpha$ has the $\alpha$-c.c. This also implies that $S_1 \cap \alpha$ is stationary in $L^{Col(\omega, <\kappa)*\mathbb{P}_\alpha}$. Moreover, in $L^{Col(\omega, <\kappa)*\mathbb{P}_\alpha}$, we have $S_1 \cap \alpha$ consists of points of cofinality of $\omega$. Thus, it makes sense to shoot a club subset of $\alpha$ with order type $\omega_1$ through $S_1 \cap \alpha$.

Let $\mathbb{P}$ be the revised limit of $((\mathbb{P}_\alpha, \dot{\mathbb{Q}}_\alpha) : \alpha \in \mathsf{Ord})$. By Lemmas 4.3, 4.4 and Theorem 4.4, we have $\mathbb{P}_\alpha$ is subcomplete for all $\alpha \in \mathsf{Ord}$. Standard arguments give us that $\mathbb{P}$ has the $\mathsf{Ord}$-c.c. Hence $\mathbb{P}$ does not add reals and $\omega_1$ is preserved. Let $G$ be $\mathbb{P}$-generic over $L[H]$. Then $L[H, G] \models \mathsf{Z}_3$. The following is stated for completeness.

**Fact 4.6**  In $L[H][G]$, if $\alpha \in S_1$, then $cf(\alpha) = \omega$; if $\alpha \in S_2$, then $cf(\alpha) = \omega_1$ and there is a club in $\alpha$ of order type $\omega_1$ contained in $S_1 \cap \alpha$.

For each $L$-cardinal $\mu > \omega_1$, we define $S_\mu = \{X \prec L_\mu : X$ is countable and $o.t.(X \cap \mu)$ is an $L$-cardinal$\}$, as being defined in the respective models of set theory which are to be considered. The following proof shows that subcomplete forcings preserve the stationarity of $S_\mu$.

**Lemma 4.7**  In $L[H, G]$, for each $L$-cardinal $\mu > \omega_1$, $S_\mu$ as defined in $L[H, G]$ is stationary.

*Proof*  Fix an $L$-cardinal $\mu > \omega_1$. Suppose $S_\mu$ is not stationary in $L[H, G]$. Then there are $p \in \mathbb{P}_\alpha$ and $\tau \in L[H]^{\mathbb{P}_\alpha}$ for some $\alpha$ such that $p \Vdash^{\mathbb{P}_\alpha}_{L[H]}$ "$\tau : [\check{\mu}]^{<\omega} \to \check{\mu}$ and there is no countable $X \subseteq \check{\mu}$ such that $X$ is closed under $\tau$ and $o.t.(X)$ is an $L$-cardinal". Let $\mu^*$ be an $L$-cardinal such that $\mu^* > \mu$. Let $\sigma : N \to L_{\mu^*}[H]$ where $N$ is countable, transitive and full, such that $\mathbb{P}_\alpha, p, \mu, \tau \in N$. Let $\sigma(\overline{\mathbb{P}}, \delta, \overline{p}, \overline{\mu}, \overline{\tau}) = \mathbb{P}_\alpha, \omega_1, p, \mu, \tau$. Let us write $N = L_\gamma[H \upharpoonright \delta]$.

Because $\kappa$ was remarkable in $L$ (cf. Definition 2.11), we may assume that $N$ was picked in such a way that $\gamma$ is an $L$-cardinal. Let $\overline{G}$ be $\overline{\mathbb{P}}$-generic over $L_\gamma[H \upharpoonright \delta]$ with $\overline{p} \in \overline{G}$. Since $\mathbb{P}_\alpha$ is subcomplete, by the definition of subcompleteness, there is $p^* \in \mathbb{P}_\alpha$ such that $p^* \leq p$ and whenever $G^*$ is $\mathbb{P}_\alpha$-generic over $L[H]$ with $p^* \in G^*$, then there is $\sigma' \in L[H][G^*]$ such that $\sigma' : L_\gamma[H \upharpoonright \delta][\overline{G}] \to L_\mu[H][G^*]$ and $\sigma'(\overline{\mathbb{P}}, \delta, \overline{p}, \overline{\mu}, \overline{\tau}) = \mathbb{P}_\alpha, \omega_1, p, \mu, \tau$.

Since $p \in G^*$, there is no countable $X \subseteq \mu$ such that $X$ is closed under $\tau^{G^*}$ and $o.t.(X)$ is an $L$-cardinal. But $ran(\sigma') \cap \mu$ is countable, closed under $\tau^{G^*}$ and $o.t.(ran(\sigma') \cap \mu) = \gamma$ is an $L$-cardinal, which leads to a contradiction.      □

*Remark 4.2* For each $L$-cardinal $\mu > \omega_1$, $S_\mu^{L[H]}$ as defined in $L[H]$ is stationary in $L[H]$; but $S_\mu^{L[H]}$ is not stationary in $L[H, G]$.[5]

We now let $\mathbb{U} = Club(Ord, S_1 \cup S_2)$. The proof of the following claim imitates the proof of Lemma 4.5.

**Lemma 4.8** $\mathbb{U}$ is $\omega_1$-distributive.

*Proof* In $L[H, G]$, $S_2$ is stationary and CH holds. Suppose $\vec{D} = (D_i : i < \omega_1)$ is $\Sigma_m$-definable sequence of open dense classes. Pick $M \prec_{\Sigma_{m+5}} V$ such that $M$ contains the parameters needed in the definition of $\vec{D}$, $M^\omega \subseteq M$ and $M \cap Ord \in S_2$.

Let $\delta = M \cap Ord$. By Fact 4.6, we may pick some $C \subseteq S_1 \cap \delta$ such that $C$ is a club in $\delta$. Now we can simultaneously build a descending sequence $(p_i : i \leq \omega_1)$ with $p_0 = p$ and a continuous tower $(M_i : i \leq \omega_1)$ of countable elementary substructures of $M$ with $M_{\omega_1} = M$ such that for all $i < \omega_1$ we have:

(1) $p_i \in M_{i+1}$,
(2) $p_{i+1} \in D_i$ and $p_{i+1}(max(dom(p_{i+1}))) > sup(M_i \cap Ord)$,
(3) $sup(M_i \cap Ord) \in C$, and
(4) if $i < \omega_1$ is a limit ordinal, then $p_i \upharpoonright max(dom(p_i)) = \bigcup_{j<i} p_j$ and hence $p_i(max(dom(p_i))) = sup(M_i \cap Ord) \in C$.

Then $p_{\omega_1} \leq p$ and $p_{\omega_1} \in \bigcap_{i<\omega_1} D_i$.      □

Let $I$ be $\mathbb{U}$-generic over $L[H, G]$, and let $C \subseteq S_1 \cup S_2$ be the club added by $I$. By Lemma 4.8, we have $L[H, G, I] \models Z_3$. As in the proof of Theorem 2.9, we can pick $B \subseteq Ord$ such that $L[H, G, I] = L[B]$ and for any $\alpha \in C$, $B$ restricted to the odd ordinals in $[\alpha, \alpha + \omega_1)$ codes a well-ordering of $min(C \backslash (\alpha + 1))$. We now reshape as follows.[6]

**Definition 4.9** Define that $p \in \mathbb{S}$ if $p : \alpha \to 2$ for some $\alpha$ and for any $\xi \leq \alpha$, $L_{\xi+1}[B \cap \xi, p \upharpoonright \xi] \models |\xi| \leq \omega_1$.

**Lemma 4.9** $\mathbb{S}$ is $\omega_1$-distributive.

*Proof* Let $(D_i : i < \omega_1)$ be a sequence of open dense subclass of $\mathbb{S}$. Let $p \in \mathbb{S}$. We want to find $p_{\omega_1}$ such that $p_{\omega_1} \in \bigcap_{i<\omega_1} D_i$ and $p_{\omega_1} \leq p$. Say $(D_i : i < \omega_1)$ is $\Sigma_m$-definable in $L[B]$ with parameters $\bar{s}$. Let $(\beta_i : i \leq \omega_1)$ be the first $\omega_1 + 1$ many $\beta$ such that $L_\beta \prec_{\Sigma_{m+5}} L[B]$ and $\omega_1 + 1 \cup \{\bar{s}\} \subseteq L_\beta[B]$. For every $i \leq \omega_1$, $(\beta_j | j < i)$

---

[5]Since $\kappa$ is remarkable in $L$, we have $S_\mu^{L[H]}$ is stationary. Take $\theta$ such that $L[H, G] \models \theta = \omega_2$. Let $A$ be a Namba sequence in $\theta$. Note that there is no structure $X \prec \langle L_\theta[H, G], A \rangle$ such that $X$ is countable and $X \in L[H]$. Thus, in $L[H, G]$, we have $S_\mu^{L[H]}$ is not stationary.

[6]In the proof of Theorem 2.9 there was no need for reshaping at this point due to (2.4).

is $\Sigma_{m+6}$-definable over $L_{\beta_i}[B]$ and hence $(\beta_j | j < i) \in L_{\beta_i+1}[B]$. Thus, for $i \le \omega_1$, $L_{\beta_i+1}[B] \models \beta_i$ is singular.

Now we define $(p_i : i \le \omega_1)$ by induction as follows. Let $p_0 = p$. Given $p_n \in \mathbb{S}$, take $p_{n+1} \in \mathbb{S}$ such that $p_{n+1} \in D_n \cap X_{n+1}$, $p_{n+1} \le p_n$ and $dom(p_{n+1}) \ge \beta_n$. Let $p_{\omega_1} = \bigcup_{i < \omega_1} p_i$. Note that $dom(p_{\omega_1}) = \beta_{\omega_1}$, $p_{\omega_1} \in \mathbb{S}$ (in fact $p_{\omega_1} \in \bigcap_{i < \omega_1} D_i$) and $p_{\omega_1} \le p$. □

By forcing with $\mathbb{S}$ over $L[H, G, I]$, we get $\overline{B} \subseteq \mathsf{Ord}$ such that for any $\alpha \in \mathsf{Ord}$, $L_{\alpha+1}[B \cap \alpha, \overline{B} \cap \alpha] \models |\alpha| \le \omega_1$. Let $E = B \oplus \overline{B}$. Note that $L[E] \models \mathsf{Z}_3$, and for any $\alpha \in \mathsf{Ord}$, $L_{\alpha+1}[E \cap \alpha] \models |\alpha| \le \omega_1$. We also have that for all $\alpha \in C$, $E$ restricted to the odd ordinals in $[\alpha, \alpha + \omega_1)$ codes a well-ordering of $\min(C \setminus (\alpha + 1))$.

By Lemmas 4.8 and 4.9, $L[H, G]$ and $L[E]$ have the same sets. Therefore, trivially, Lemma 4.7 is still true with $L[E]$ replacing $L[H, G]$.

Exactly as in the proof of Theorem 2.9 we can do almost disjoint forcing to add $A \subseteq \omega_1$ to code E. Note that $L[E][A] = L[A]$ and the forcing we use to add $A$ is countably closed and $\mathsf{Ord}$-c.c. Since $L[E] \models \mathsf{Z}_3$, $L[A] \models \mathsf{Z}_3$. By the countable closure, Lemma 4.7 is still true with $L[A]$ replacing $L[H, G]$.

By the same argument as in Theorem 2.9 we can show that if $\alpha > \omega_1$ is $A$-admissible then $\alpha \in C$, and hence $L \models \varphi(\alpha)$. By our hypothesis on $\kappa$, we have $L \models \varphi(\kappa)$ and if $\alpha \ge \omega_1$ is $A$-admissible then $L \models \varphi(\alpha)$. Now we do reshaping over $L[A]$ as follows.

**Definition 4.10** Define $p \in \mathbb{T}$ iff $p : \alpha \to 2$ for some $\alpha < \omega_1$ and for any $\xi \le \alpha$, there is $\gamma$ such that $L_\gamma[A \cap \xi, p \restriction \xi] \models$ "$\xi$ is countable" and if $\lambda \in [\xi, \gamma]$ is $(A \cap \xi)$-admissible, then $L \models \varphi(\lambda)$.

**Lemma 4.10** $\mathbb{T}$ is $\omega$-distributive.

*Proof* Recall that for each $L$-cardinal $\mu > \omega_1$, we defined $S_\mu = \{X \prec L_\mu : X$ is countable and $o.t.(X \cap \mu)$ is an $L$-cardinal$\}$. We shall use the fact that in $L[A]$, $S_\mu$ as defined in $L[A]$ is stationary. In fact, essentially the same argument as in the proof of Lemma 2.3 shows that $\mathbb{T}$ is $\omega$-distributive. In the following we only point out the place we use that $\varphi$ is $\Sigma_2$ in our argument.

Let $p \in \mathbb{T}$ and $\overrightarrow{D} = (D_n : n \in \omega)$ be a sequence of open dense sets. Pick large enough $L$-cardinal $\mu$ such that $\overrightarrow{D} \in L_\mu[A]$ and $L_\mu[A] \models$ "if $\alpha \ge \omega_1$ is $A$-admissible, then $L \models \varphi(\alpha)$". Since $S_\mu$ is stationary, we can pick X such that $\pi : L_{\overline{\mu}}[A \cap \delta] \cong X \prec L_\mu[A]$, $|X| = \omega$, $\{p, \mathbb{T}, A, \overrightarrow{D}, \omega_1, \nu\} \subseteq X$ and $\overline{\mu}$ is an $L$-cardinal where $\pi(\delta) = \omega_1 (\delta = X \cap \omega_1)$. Note that by elementarity, we have $L_{\overline{\mu}}[A \cap \delta] \models$ "if $\alpha \ge \delta$ is $A \cap \delta$-admissible, then $L \models \varphi(\alpha)$". Suppose $\alpha \in [\delta, \overline{\mu})$ is $A \cap \delta$-admissible. Then $L_{\overline{\mu}} \models \varphi(\alpha)$. Since $\overline{\mu}$ is an $L$-cardinal and $\varphi$ is $\Sigma_2$, we have $L \models \varphi(\alpha)$. The rest of the arguments are the same as in the proof of Lemma 2.3. □

Using Lemma 4.6, a simple variant of the previous proof also shows the following.

**Lemma 4.11** $\{\alpha < \kappa : L \models \varphi(\alpha)\}$ is stationary in $L[A]^{\mathbb{T}}$.

Forcing with $\mathbb{T}$ adds $F : \omega_1 \to 2$ such that for any $\alpha < \omega_1$, there exists $\gamma$ such that $L_\gamma[A \cap \alpha, F \upharpoonright \alpha] \models$ "$\alpha$ is countable" and every $(A \cap \alpha)$-admissible $\lambda \in [\alpha, \gamma]$ satisfies that $L \models \varphi(\lambda)$. Using Lemma 4.6, we may force over $L[A, F]$ and shoot a club $C^*$ through $\{\alpha < \kappa : L \models \varphi(\alpha)\}$ in the standard way. Let $D = A \oplus F \oplus C^*$. We may assume that for $\lambda \in C^*$, $D$ restricted to odd ordinals in $[\lambda, \lambda + \omega)$ codes a well-ordering of $\min(C^* \setminus (\lambda + 1))$. Since $\mathbb{T}$ and the club shooting adding $C^*$ are $\omega$-distributive, it is easy to see that $L[D] \models Z_3$.

Now we work in $L[D]$. Do almost disjoint forcing to code $D$ by a real $x$. This forcing is c.c.c. Note that $L[D][x] = L[x]$ and $L[x] \models Z_3$.

Now we work in $L[x]$. Suppose $\alpha$ is $x$-admissible. We show that $L \models \varphi(\alpha)$. If $\alpha \geq \omega_1$, then $\alpha$ is also $A$-admissible and hence $L \models \varphi(\alpha)$. Now we assume that $\alpha < \omega_1$ and $L \nvDash \varphi(\alpha)$. Then $\alpha \notin C^*$. Let $\lambda$ be the largest element of $C^*$ such that $\lambda < \alpha$ and $\bar{\lambda} = \min(C \setminus (\alpha + 1)) > \alpha$. For every $\xi < \omega_1$, let $\xi^* > \xi$ be least such that $L_{\xi^*}[A \cap \xi, F \upharpoonright \xi] \models \xi$ is countable. By the properties of $F$, we have every $(D \cap \xi)$-admissible $\lambda' \in [\xi, \xi^*]$ satisfies $L \models \varphi(\lambda')$.

Case 1: For all $\xi < \lambda + \omega$, $\xi^* < \alpha$. Then $D \cap (\lambda + \omega)$ can be computed inside $L_\alpha[x]$. But then, as $\alpha$ is $x$-admissible, the ordinal coded by $D$ restricted to the odd ordinals in $[\lambda, \lambda + \omega)$, namely $\bar{\lambda}$, is in $L_\alpha[x]$ and hence $\bar{\lambda} < \alpha$, which leads to a contradiction.

Case 2: Not Case 1. Let $\xi < \lambda + \omega$ be least such that $\xi^* \geq \alpha$. Then $D \cap \xi$ can be computed inside $L_\alpha[x]$. Since $\alpha$ is $x$-admissible, $\alpha$ is $(D \cap \xi)$-admissible. But all $(D \cap \xi)$-admissible $\lambda' \in [\xi, \xi^*]$ satisfy $L \models \varphi(\lambda')$. Hence $L \models \varphi(\alpha)$ since $\xi < \alpha \leq \xi^*$, which leads to a contradiction.

We have shown that $L[x] \models Z_3 + HP(\varphi)$. $\quad\square$

**Theorem 4.7** *The following two theories are equi-consistent:*

*(1)* $ZFC+$ *there exists a remarkable cardinal* $\kappa$ *with* $\varphi(\kappa) +$ Ord *is* $2$-$\varphi$-*Mahlo.*
*(2)* $Z_3 + HP(\varphi)$.

*Proof* Follows from Proposition 4.1 and Theorem 4.5. $\quad\square$

**Corollary 4.2** $Z_3 + HP(\varphi)$ *does not imply that* $0^\sharp$ *exists.*

By Corollary 2.1 or Theorem 2.12, $Z_4 + HP(\varphi)$ implies that $0^\sharp$ exists. As a corollary, $Z_4$ is the minimal system of higher-order arithmetic to show that $HP$, $HP(\varphi)$ and $0^\sharp$ exists are equivalent with each other.

# References

1. Cheng, Y., Schindler, R.: Harrington's Principle in higher-order arithmetic. J. Symb. Log. **80**(02), 477–489 (2015)
2. Harrington, L.A., Baumgartner, J.E., Kleinberg, E.M.: Adding a closed unbounded set. J. Symb. Log. **41**, 481–482 (1976)

3. Jensen, R.B.: Lecture note on subcomplete forcing and L-forcing. Handwritten notes. https://www.mathematik.hu-berlin.de/~raesch/org/jensen.html
4. Shelah, S.: Proper and Improper Forcing. Perspectives in Math. Logic. Springer (1998)
5. Cummings, J.: Iterated forcing and elementary embeddings. In: Foreman, M., Kanamori, A. (Eds.) Chapter 12 in Handbook of Set Theory. Springer, Berlin (2010)
6. Jensen, R.B.: Iteration Theorems for Subcomplete and Related Forcings. Handwritten notes. https://www.mathematik.hu-berlin.de/~raesch/org/jensen.html

# Chapter 5
# Forcing a Model of Harrington's Principle Without Reshaping

**Abstract** In this chapter, we establish the following main result: assuming there exists a remarkable cardinal with a weakly inaccessible cardinal above it, we can force a set model of $Z_3 + HP$ via set forcing *without* the use of the reshaping technique.

## 5.1 Introduction

In Chap. 2, we have proved that $Z_3 + HP$ is equiconsistent with "ZFC + there exists a remarkable cardinal" via class forcing. As an easy corollary of this theorem, we can force a set model of $Z_3 + HP$ via set forcing assuming there exists a remarkable cardinal. However, the proof in Sect. 2.3 uses the reshaping technique.[1] In this chapter, I force a model of $Z_3 + HP$ via set forcing without the use of the reshaping technique. The method in this chapter is totally different from Chap. 2 and has its own merit and power.

The structure of this chapter is as follows. In Sect. 5.2, I introduce the notion of strong reflecting property for $L$-cardinals to be used in the proof of Theorem 5.1. In Sect. 5.3, I introduce Baumgartner's forcing to be used in Sect. 5.6. The proof of Theorem 5.1 consists of four steps in resp. Sections 5.4–5.7. I first give an outline of the proof of Theorem 5.1 in Sect. 5.4.

## 5.2 The Notion of Strong Reflecting Property for $L$-Cardinals

The notion of *strong reflecting property* for $L$-cardinals was first introduced in [2]. In this section, I prove some basic properties of the strong reflecting property for $L$-cardinals to be used in our proof of Theorem 5.1. In Chap. 6, I will develop the full

---

[1]For the original version of reshaping forcing, I refer to [1, Sect. 1.3].

© The Author(s), under exclusive license to Springer Nature Singapore Pte Ltd. 2019
Y. Cheng, *Incompleteness for Higher-Order Arithmetic*, SpringerBriefs
in Mathematics, https://doi.org/10.1007/978-981-13-9949-7_5

theory of the strong reflecting property for $L$-cardinals and examine more properties not covered in this section.

**Definition 5.1** Let $\gamma$ be an $L$-cardinal. If $\gamma \geq \omega_1$, we define that $\gamma$ has the strong reflecting property if for some regular cardinal $\kappa > \gamma$, for any $X \in [H_\kappa]^\omega$ if $X \prec H_\kappa$ and $\gamma \in X$, then $\overline{\gamma}$ is an $L$-cardinal. If $\gamma < \omega_1$, we define that $\gamma$ has the strong reflecting property if $\gamma = \overline{\gamma}$.

Note that under $V = L$, $H_\eta = L_\eta$ for any $L$-cardinal $\eta$. In this chapter, I often use $H_\eta$ and $L_\eta$ interchangeably. Throughout this chapter whenever I write $X \prec H_\kappa$ and $\gamma \in X$, $\overline{\gamma}$ always denotes the image of $\gamma$ under the transitive collapse of $X$.

**Definition 5.2** If $X \subseteq \gamma$ and $F : \gamma^{<\omega} \to \gamma$, we define that $X$ is closed under $F$ if $F``X^{<\omega} \subseteq X$.

**Proposition 5.1** *For an $L$-cardinal $\gamma \geq \omega_1$, the following are equivalent:*

(a) *$\gamma$ has the strong reflecting property;*
(b) *There exists $F : \gamma^{<\omega} \to \gamma$ such that if $X \subseteq \gamma$ is countable and closed under $F$, then $o.t.(X)$ is an $L$-cardinal;*
(c) *For any regular cardinal $\kappa > \gamma$, for any $X \in [H_\kappa]^\omega$ if $X \prec H_\kappa$ and $\gamma \in X$, then $\overline{\gamma}$ is an $L$-cardinal.*

*Proof* $(a) \Rightarrow (b)$: Let $\kappa > \gamma$ be the witness regular cardinal for $(a)$. Let $Z = \{X \mid X \prec H_\kappa, |X| = \omega, \gamma \in X$ and $\overline{\gamma}$ is an $L$-cardinal$\}$. Then $Z \upharpoonright \gamma = \{X \cap \gamma \mid X \in Z\}$ contains a club $E$ in $[\gamma]^\omega$. Thus, there exists $F : \gamma^{<\omega} \to \gamma$ such that if $X \subseteq \gamma$ is countable and closed under $F$, then $X \in E$. Suppose $X \subseteq \gamma$ is countable and closed under $F$. We show that $o.t.(X)$ is an $L$-cardinal. Since $X \in E$, we have $X = Y \cap \gamma$ for some $Y \in Z$ and hence $\overline{\gamma}$ is an $L$-cardinal. Thus, $o.t.(X) = o.t.(Y \cap \gamma) = \overline{\gamma}$ is an $L$-cardinal.

$(b) \Rightarrow (c)$: Suppose $\kappa > \gamma$ is regular, $X \prec H_\kappa$, $|X| = \omega$ and $\gamma \in X$. We show that $\overline{\gamma}$ is an $L$-cardinal. By (b), take $F \in X$ such that in $X$, $F : \gamma^{<\omega} \to \gamma$ has the property:

$$\text{if } X \subseteq \gamma \text{ is countable and closed under } F, \text{ then } o.t.(X) \text{ is an } L\text{-cardinal.} \quad (5.1)$$

Since $X \cap \gamma$ is closed under $F$, by (5.1), $o.t.(X \cap \gamma)$ is an $L$-cardinal. But $\overline{\gamma} = o.t.(X \cap \gamma)$. Thus, $\overline{\gamma}$ is an $L$-cardinal. $\square$

**Proposition 5.2** *Suppose $\gamma \geq \omega_1$ is an $L$-cardinal and $|\gamma| = \omega_1$. Then the following are equivalent:*

(a) *$\gamma$ has the strong reflecting property;*
(b) *For any bijection $\pi : \omega_1 \to \gamma$, there exists a club $D \subseteq \omega_1$ such that for any $\theta \in D$, $o.t.(\{\pi(\alpha) \mid \alpha < \theta\})$ is an $L$-cardinal;*
(c) *For some bijection $\pi : \omega_1 \to \gamma$, there exists a club $D \subseteq \omega_1$ such that for any $\theta \in D$, $o.t.(\{\pi(\alpha) \mid \alpha < \theta\})$ is an $L$-cardinal.*

*Proof* $(a) \Rightarrow (b)$: Let $\kappa > \gamma$ be the regular cardinal that witnesses the strong reflecting property of $\gamma$. Suppose $\pi : \omega_1 \to \gamma$ is a bijection. Let $E = \{X \cap \omega_1 \mid X \prec H_\kappa, |X| = \omega, \pi \in X$ and $\gamma \in X\}$. Then $E$ contains a club $D$ in $\omega_1$. Let $\beta \in D$. Then $\beta = X \cap \omega_1$ for some $X$ such that $\pi \in X$, $X \prec H_\kappa$, $|X| = \omega$ and $\gamma \in X$. Note that $\overline{\gamma} = o.t.(\{\pi(\alpha) \mid \alpha < X \cap \omega_1\})$. Thus $o.t.(\{\pi(\alpha) \mid \alpha < \beta\}) = \overline{\gamma}$ is an $L$-cardinal.

$(c) \Rightarrow (a)$: Let $\kappa > \gamma$ be a regular cardinal with $\kappa \geq (2^{\omega_1})^+$. Suppose $X \prec H_\kappa$, $|X| = \omega$ and $\gamma \in X$. We show that $\overline{\gamma}$ is an $L$-cardinal. By $(c)$, take $\pi, D \in X$ such that $\pi : \omega_1 \to \gamma$ is a bijection and $D \subseteq \omega_1$ is the witness club for $\pi$ in $(c)$. Since $D$ is unbounded in $X \cap \omega_1$, we have $X \cap \omega_1 \in D$. Note that $\overline{\gamma} = o.t.(\{\pi(\alpha) \mid \alpha \in X \cap \omega_1\})$. Thus, $\overline{\gamma}$ is an $L$-cardinal.  □

Let $(i)^*$, $(ii)^*$ and $(iii)^*$ denote the statements of Propositions 5.1(a), 5.1(c) and 5.2(b) with "is an $L$-cardinal" replaced with "is not an $L$-cardinal", respectively. The following corollary is an observation from proofs of Propositions 5.1 and 5.2.

**Corollary 5.1**  *Suppose $\gamma \geq \omega_1$ is an $L$-cardinal and $|\gamma| = \omega_1$. Then $(i)^*$, $(ii)^*$ and $(iii)^*$ are equivalent.*

**Proposition 5.3**  *Suppose $\gamma \geq \omega_1$ is an $L$-cardinal. The statement "$\gamma$ has the strong reflecting property" is upward absolute.*

*Proof* Suppose $M \subseteq N$ are inner models and $M \models$ "$\gamma \geq \omega_1$ has the strong reflecting property". We show that $N \models$ "$\gamma$ has the strong reflecting property".

By Proposition 5.1, in $M$, there exists $F : \gamma^{<\omega} \to \gamma$ such that (5.1) holds. If $\gamma$ is countable in $N$, by definition, $\gamma$ has the strong reflecting property in $N$. Assume that $N \models \gamma$ is uncountable. By Proposition 5.1, it suffices to show that in $N$, (5.1) holds.

Suppose not. Then in $N$, there exists $\overline{\gamma} < \omega_1$ such that $\overline{\gamma}$ is not an $L$-cardinal and there exists an order preserving $j : \overline{\gamma} \to \gamma$ such that $ran(j)$ is closed under $F$. Thus, in $N$, there exists $e : \omega \to L_{\omega_1^N}$ and $\gamma' \in e``\omega$ such that $e``\omega \prec L_{\omega_1^N}$, $L_{\omega_1^N} \models$ "$\gamma'$ is not an $L$-cardinal" and there exists an order preserving $j' : o.t.(e``\omega \cap \gamma') \to \gamma$ such that $ran(j')$ is closed under $F$.

Let $\langle \varphi_i \mid i \in \omega \rangle$ be a recursive enumeration of formulas with infinite repetitions. We assume that for $i \in \omega$, $\varphi_i$ has free variables among $x_0, \ldots, x_{i+1}$. Thus, in $N$ there exist $e : \omega \to L_{\omega_1^N}$, $\pi : \omega \to \gamma$ and $\gamma^* \in e``\omega$ such that

(1) for any $i \in \omega$, if there exists $a \in L_{\omega_1^N}$ such that $L_{\omega_1^N} \models \varphi_i[a, e(0), \ldots, e(i)]$, then $L_{\omega_1^N} \models \varphi_i[e(2i + 1), e(0), \ldots, e(i)]$;
(2) $ran(\pi)$ is closed under $F$;
(3) $L_{\omega_1^N} \models \gamma^*$ is not an $L$-cardinal; and
(4) for $i \in \omega$, if $e(i) \notin \gamma^*$, then $\pi(i) = 0$; for $i < j \in \omega$, if $e(i), e(j) \in \gamma^*$, then $\pi(i) < \pi(j) \Leftrightarrow e(i) < e(j)$ and $\pi(i) = \pi(j) \Leftrightarrow e(i) = e(j)$.

In $N$, let $T = \{(e \restriction n, \pi \restriction n) : e$ and $\pi$ have properties (1)–(4)$\}$. $T$ is a tree and from (1)–(4), by absoluteness, $T \in M$. Since in $N$, there exists $(e, \pi)$ satisfying (1)–(4), $T$ has an infinite branch in $N$. By absoluteness, $T$ has an infinite branch in $M$ and such a branch corresponds to the existence of $(e, \pi)$ with properties (1)–(4) in $M$. Thus, in $M$ there exists $X \subseteq \gamma$ such that $X$ is countable, closed under $F$ and $o.t.(X)$ is not an $L$-cardinal which contradicts (5.1).  □

## 5.3  Baumgartner's Forcing

Our proof of Theorem 5.1 uses Baumgartner's forcing in Sect. 5.6. In this section, I introduce Baumgartner's forcing and prove some properties of Baumgartner's forcing in $Z_3$. In this section, we assume that $S$ is a stationary subset of $\omega_1$.

**Definition 5.3** (*Baumgartner's forcing*, [3]) Define $\mathbb{P}_S^B = \{f : dom(f) \to S \mid dom(f) \subseteq \omega_1$ is finite and $\exists \alpha > max(dom(f)) \exists g : \alpha \to S(g$ is continuous, increasing and $g \restriction dom(f) = f)\}$. For $f, g \in \mathbb{P}_S^B$, define $g \leq f$ if $f \subseteq g$.

Note that the following are equivalent:

(1) $f \in \mathbb{P}_S^B$;
(2) $dom(f) \subseteq \omega_1$ is finite and there exists $g : max(dom(f)) + 1 \to S$ such that $g$ is continuous, increasing and $g \restriction dom(f) = f$;
(3) $dom(f) \subseteq \omega_1$ is finite and there exists $C \subseteq S$ such that $C$ is closed, $o.t.(C) = max(dom(f)) + 1$ and for any $\beta \in dom(f)$, $f(\beta)$ is the $\beta$-th element of $C$.

Note that if $F : \omega_1 \to S$ is increasing and continuous, then $ran(F) \subseteq S$ is a club on $\omega_1$. If $C \subseteq S$ is a club on $\omega_1$, then $F : \omega_1 \to S$ is increasing and continuous where $F(\alpha) =$ the $\alpha$-th element of $C$.

Let $G$ be $\mathbb{P}_S^B$-generic over $V$. Define $F_G = \bigcup\{f \mid f \in G\}$. From the definition of $\mathbb{P}_S^B$, it is not difficult to check that: (1) for any $f \in \mathbb{P}_S^B$ and all $\alpha < \omega_1$, there exists $g \in \mathbb{P}_S^B$ such that $g \leq f$ and $\alpha \in dom(g)$; (2) for any $f \in \mathbb{P}_S^B$ and any $\alpha \in dom(f)$, if $\alpha$ is a limit ordinal, then for any $\eta < f(\alpha)$, there exist $g \in \mathbb{P}_S^B$ and $\beta < \alpha$ such that $g \leq f$, $\beta \in dom(g)$ and $g(\beta) > \eta$. Thus, $F_G : \omega_1 \to S$ is increasing and continuous. Let $C = ran(F_G)$. Then $C \subseteq S$ is a club on $\omega_1$. Let $D = \{\alpha \mid \alpha$ is a limit point of $C\}$. Note that $D \subseteq C$ is a club on $\omega_1$.

For $f \in \mathbb{P}_S^B$, define $(\mathbb{P}_S^B)_f = \{g \in \mathbb{P}_S^B \mid g \leq f$ and $max(dom(g)) = max (dom(f))\}$. Note that $|(\mathbb{P}_S^B)_f| = \omega$ for $f \in P_S^B$.

**Proposition 5.4** ($Z_3$) *Suppose* $f \in \mathbb{P}_S^B$. *Then* $f \Vdash_{\mathbb{P}_S^B} \dot{G} \cap (\mathbb{P}_S^B)_f$ *is* $(\mathbb{P}_S^B)_f$-*generic over* $V$.

*Proof* Assume $h \in \mathbb{P}_S^B$, $h \leq f$ and $D$ is a dense subset of $(\mathbb{P}_S^B)_f$. It suffices to show that there is $p \in D$ such that $h \cup p \in \mathbb{P}_S^B$. Let $max(dom(f)) = \beta$. Then $h \restriction (\beta + 1) \in (\mathbb{P}_S^B)_f$. Take $p \in D$ such that $p \leq h \restriction (\beta + 1)$. We show that $h \cup p \in \mathbb{P}_S^B$.

Let $\alpha = max(dom(h))$. Since $h \in \mathbb{P}_S^B$, there exists $E \subseteq S$ such that $E$ is closed, $o.t.(E) = \alpha + 1$ and for any $\gamma \in dom(h)$, $h(\gamma)$ is the $\gamma$-th element of $E$. Since $p \in (\mathbb{P}_S^B)_f$, we have $max(dom(p)) = \beta$. Let $F \subseteq S$ be closed such that $o.t.(F) = \beta + 1$ and for any $\gamma \in dom(p)$, $p(\gamma)$ is the $\gamma$-th element of $F$. Note that $h(\beta) = f(\beta) = p(\beta)$. Let $C = \{\gamma \in E \mid \gamma \geq h(\beta) = p(\beta)\} \cup F$. $C \subseteq S$ is closed. Since $p \leq h \restriction (\beta + 1)$, $o.t.(C) = \alpha + 1$. For any $\gamma \in dom(h \cup p)$, $(h \cup p)(\gamma)$ is the $\gamma$-th element of $C$. Thus, $h \cup p \in \mathbb{P}_S^B$. $\qquad\square$

**Definition 5.4** A limit ordinal $\gamma$ is *indecomposable* if there exist no $\alpha < \gamma$ and $\beta < \gamma$ such that $\alpha + \beta = \gamma$.

Note that (1) a limit ordinal $\gamma$ is indecomposable if and only if $\alpha + \gamma = \gamma$ for any $\alpha < \gamma$ if and only if $\gamma = \omega^\alpha$ for some $\alpha$ (this is ordinal exponentiation); (2) if $\gamma$ is indecomposable, then for any $\alpha < \gamma$, $o.t.(\{\beta \mid \alpha \le \beta < \gamma\}) = \gamma$. For $\eta < \omega_1$, define $\mathbb{P}_S^B \upharpoonright \eta = \{f \in \mathbb{P}_S^B \mid (dom(f) \cup ran(f)) \subseteq \eta\}$.

**Lemma 5.1** ($\mathsf{Z}_3$) *Suppose $\eta < \omega_1$ is indecomposable and $f \in \mathbb{P}_S^B$ where $f = \{(\eta, \eta)\}$. Then*

$$(\mathbb{P}_S^B)_f = \{g \cup \{(\eta, \eta)\} \mid g \in \mathbb{P}_S^B \upharpoonright \eta\}.$$

*Proof* $\subseteq$ is trivial. Fix $g \in \mathbb{P}_S^B \upharpoonright \eta$. We show that $g \cup \{(\eta, \eta)\} \in \mathbb{P}_S^B$. It suffices to show that there exists $H : \eta + 1 \to S \cap (\eta + 1)$ such that

$$H \text{ is increasing and continuous, } H \text{ extends } g \text{ and } H(\eta) = \eta. \tag{5.2}$$

Let $\xi = \max(dom(g))$. Let $F : \xi + 1 \to S \cap (g(\xi) + 1)$ be the witness function for $g \in \mathbb{P}_S^B$ (i.e. $F$ is increasing, continuous and extends $g$). Let $E : \eta + 1 \to S \cap (\eta + 1)$ be the witness function for $f \in \mathbb{P}_S^B$ (i.e. $E$ is increasing, continuous and $E(\eta) = \eta$). Let $C = ran(E) \setminus (g(\xi) + 1)$. Since $\eta$ is indecomposable and $g(\xi) < \eta$, $o.t.(C) = o.t.((\eta + 1) \setminus (g(\xi) + 1)) = \eta + 1$. Let $\pi : \eta + 1 \to C$ be an increasing continuous enumeration of $C$. Define $H : \eta + 1 \to S \cap (\eta + 1)$ by $H \upharpoonright (\xi + 1) = F$ and $H(\xi + 1 + \alpha) = \pi(\alpha)$ for any $\alpha \le \eta$. It is easy to check that $H$ satisfies (5.2). $\qquad\square$

**Proposition 5.5** (Baumgartner, [3], $\mathsf{Z}_3$) $\mathbb{P}_S^B$ *preserves $\omega_1$.*[2]

*Proof* It suffices to show that $\mathbb{P}_S^B$ adds only Cohen reals. Let $\tau$ be the term for a new real in the forcing language. Define

$$R_\tau = \{(1, i, f) \mid i \in \omega \wedge f \in \mathbb{P}_S^B \wedge f \Vdash i \in \tau\} \cup \{(0, i, f) \mid i \in \omega \wedge f \in \mathbb{P}_S^B \wedge f \Vdash i \notin \tau\}.$$

Take $X \prec H_{\omega_1}$ such that $|X| = \omega$ and $X \cap \omega_1 \in S$. Such $X$ exists since $S$ is stationary. Let $\eta = X \cap \omega_1$ and $f = \{(\eta, \eta)\}$. Suppose $G$ is $\mathbb{P}_S^B$-generic over $V$ and $f \in G$. We show that $\tau^G$ is a Cohen real.

Note that $\mathbb{P}_S^B \cap X = \mathbb{P}_S^B \upharpoonright \eta$. By Lemma 5.1, $\mathbb{P}_S^B \cap X \cong (\mathbb{P}_S^B)_f$. Since $f \in \mathbb{P}_S^B$, by Proposition 5.4, $f \Vdash \dot{G} \cap X$ is $(\mathbb{P}_S^B \cap X)$-generic over $V$. Thus, $G \cap X$ is $\mathbb{P}_S^B \cap X$-generic over $V$. Note that $\tau^G = \{i \in \omega \mid \exists f \in G \cap X((1, i, f) \in R_\tau)\}$. Thus, $\tau^G \in V[G \cap X]$. Since $G \cap X$ is countable, $\tau^G$ is a Cohen real. $\qquad\square$

Note that $|\mathbb{P}_S^B| = \omega_1$, even if we are not assuming **CH**. Since $\mathbb{P}_S^B$ has $\omega_2$-c.c. and preserves $\omega_1$, $\mathbb{P}_S^B$ preserves all cardinals.

Suppose $G$ is $\mathbb{P}_S^B$-generic over $V$ and $C \subseteq S$ is the new club in $V[G]$. Then for cofinally many $\alpha < \omega_1$, $C \cap \alpha \notin V$ and $C \cap \alpha$ codes a Cohen real. Thus, $\mathbb{P}_S^B$ is not $\omega$-distributive.

**Corollary 5.2** ($\mathsf{Z}_3$) $\mathbb{P}_S^B$ *has the following properties:*

---

[2]We observe that Baumgartner's proof of this fact in [3] can be done in $\mathsf{Z}_3$.

(i) $\mathbb{P}_S^B$ *is not proper;*
(ii) $\mathbb{P}_S^B$ *is not $\omega_1$-closed;*
(iii) $\mathbb{P}_S^B$ *is not $\omega$-distributive and adds new reals;*
(iv) $\mathbb{P}_S^B$ *adds only Cohen reals;*
(v) $\mathbb{P}_S^B$ *preserves $\omega_1$.*

## 5.4  The First Step

In the rest of this chapter, I will prove the following main theorem.

**Theorem 5.1** *Assuming there exists a remarkable cardinal with a weakly inaccessible cardinal above it, we can force a model of $Z_3 + HP$ via set forcing the use of the reshaping technique.*

As a corollary of Theorem 5.1, $Z_3 + HP$ does not imply that $0^\sharp$ exists. The history of the main theorem in this chapter is as follows: The theorem "$Z_3 + HP$ does not imply that $0^\sharp$ exists" was first proved in [2] via set forcing without the use of the reshaping technique. The proof of Theorem 5.1 in this book is based on [2] and I improve the presentation in [2] by computing the upper bound of the large cardinal hypothesis used in Sect. 5.4 via the notion of remarkable cardinal which is much weaker than the large cardinal hypothesis used in [2].

In the following, I first give an outline of the proof of Theorem 5.1. In Sects. 5.4–5.7, our proof of Theorem 5.1 is divided into four steps. In Sect. 5.4, I force over $L$ to get a club in $\omega_2$ of $L$-cardinals with the strong reflecting property (cf. Definition 5.1). This is necessary to show in the second step that (5.5) holds. In Sect. 5.5, I find some $B_0 \subseteq \omega_2$ and $A \subseteq \omega_1$ such that (5.5) holds in $L[B_0, A]$. The definition of $S$ is motivated by (5.5) which is necessary to show that $S$ as defined in (5.13) contains a club in $\omega_1$ and hence is stationary.

In Sect. 5.6, I shoot a club $C$ through $S$ via Baumgartner's forcing such that (5.28) holds. We use (5.28) to define the almost disjoint system and show that the generic real via almost disjoint forcing satisfies HP. In Sect. 5.7, I use properties of $Lim(C)$ (Lemmas 5.9 and 5.12) to define the almost disjoint system on $\omega$ and some $B^* \subseteq \omega_1$. Then I do almost disjoint forcing to code $B^*$ by a real $x$. Finally, I use properties of $Lim(C)$ ((5.26), (5.27) and (5.28)) to show that $x$ is the witness real for HP.

In the first step, assuming there exists a remarkable cardinal with a weakly inaccessible cardinal above it, I force over $L$ to get a club in $\omega_2$ of $L$-cardinals with the strong reflecting property.

In the following, we work in $L$. Let $\kappa$ be a remarkable cardinal and $\lambda > \kappa$ be an inaccessible cardinal. Suppose $\overline{G}$ is $Col(\omega, < \kappa)$-generic over $L$ and $G$ is $Col(\omega, < \kappa) * Col(\kappa, < \lambda)$-generic over $L$. Now we work in $L[G]$. Define $K = \{\gamma \mid \omega_1 \leq \gamma < \omega_2$ and $\gamma$ is an $L$-cardinal$\}$.

**Definition 5.5** For $\gamma \in K$, we say $\gamma$ is *weakly reflecting* if for some bijection $\pi : \omega_1 \to \gamma$, there exists stationary $D \subseteq \omega_1$ such that for any $\theta \in D$, $o.t.(\{\pi(\alpha) \mid \alpha < \theta\})$ is an $L$-cardinal.

**Proposition 5.6** $L[G] \models$ "all $\gamma \in K$ are weakly reflecting".

*Proof* We work in $L[G]$. Suppose $\gamma \in K$ is a counterexample and $\theta > \gamma$ is a regular cardinal. Since $\kappa$ is remarkable in $L$, by Lemma 2.2, we have $L[\overline{G}] \models$ "$\{X : X \prec H_\theta \wedge |X| = \omega \wedge \gamma \in X \wedge \overline{\gamma}$ is an $L$-cardinal$\}$ is stationary". Note that the property "$X \prec H_\theta \wedge |X| = \omega \wedge \gamma \in X \wedge \overline{\gamma}$ is an $L$-cardinal" is absolute between $L[\overline{G}]$ and $L[G]$. Thus, by absoluteness, in $L[G]$,

$$\exists X (X \prec H_\theta \wedge |X| = \omega \wedge \gamma \in X \wedge \overline{\gamma} \text{ is an } L\text{-cardinal}). \tag{5.3}$$

Since $\gamma$ is not weakly reflecting, $(iii)^*$ in Corollary 5.1 holds and hence, by Corollary 5.1, $(ii)^*$ holds which contradicts (5.3). $\square$

By the above, $K$ is a club in $\omega_2$ of weakly reflecting $L$-cardinals. For $\gamma \in K$, by Proposition 5.6, there exist a bijection $\pi : \omega_1 \leftrightarrow \gamma$ and a stationary set $S \subseteq \omega_1$ such that for any $\theta \in S$, $o.t.(\{\pi(\alpha) \mid \alpha < \theta\})$ is an $L$-cardinal (let $\pi_\gamma$ and $S_\gamma$ be such $\pi$ and $S$). Then $S_\gamma$ is stationary for $\gamma \in K$.

**Definition 5.6** ([4]) Assume $\kappa$ is a regular cardinal and $\{\mathbb{P}_i : i \in I\}$ is a collection of partial orders. The $\kappa$-*product* of $\{\mathbb{P}_i : i \in I\}$ is defined as $\{p : dom(p) = I \wedge \forall i \in I(p(i) \in \mathbb{P}_i) \wedge |suppt(p)| < \kappa\}$ where $suppt(p) = \{i \in I : p(i) \neq 1_{\mathbb{P}_i}\}$.

**Fact 5.2** ([4]) *Assume $\kappa^{<\kappa} = \kappa$. If $|\mathbb{P}_i| \leq \kappa$ for every $i \in I$, then the $\kappa$-product of $\mathbb{P}_i$ is $\kappa^+$-c.c.*

Let $\mathbb{P}$ be the $\omega_1$-product of $\{\mathbb{P}_\gamma : \gamma \in K\}$ where $\mathbb{P}_\gamma$ is Harrington's forcing which shoots a club through $S_\gamma$.[3] Since **CH** holds in $L[G]$, we have $|\mathbb{P}_\gamma| = \omega_1$ for $\gamma \in K$.

In $L[G]$, $\omega_1^{<\omega_1} = \omega_1$. By Fact 5.2, $\mathbb{P}$ is $\omega_2$-c.c. For $\gamma \in K$, $\mathbb{P}_\gamma$ is $\omega$-distributive and hence preserves $\omega_1$. The proof of the following lemma imitates [6, Lemma 2.4].

**Lemma 5.2** $\mathbb{P}$ *is $\omega$-distributive.*

*Proof* For $\gamma \in K$, we may view $\mathbb{P}_\gamma$ as the set of all strictly increasing and continuous sequences $(\eta_i : i \leq \alpha)$ of countable successor length consisting of elements of $S_\gamma$. For $p \in \mathbb{P}$, we may write $p = \{(\eta_i^\lambda(p) : i \leq \alpha_\lambda(p)) | \lambda \in suppt(p)\}$. Let $\overrightarrow{D} = (D_n : n \in \omega)$ be a sequence of dense open subsets of $\mathbb{P}$. Let $p \in \mathbb{P}$. Pick some $Y \prec H_{\omega_3}$ such that $\omega_1 \cup \{p, \mathbb{P}, \overrightarrow{D}\} \subseteq Y$, $Y \cap \omega_2 < \omega_2$ and $Y$ is of cardinality $\omega_1$. Let $\gamma = Y \cap \omega_2$. Then $\gamma$ is an $L$-cardinal and $\gamma \in K$. Since $S_\gamma$ is stationary, we may pick some countable $X \prec H_{\omega_3}$ such that $\{p, \mathbb{P}, \overrightarrow{D}, Y, \gamma\} \subseteq X$ and $X \cap \gamma \in S_\gamma$. Then we have $\{p, \mathbb{P}, \overrightarrow{D}\} \subseteq X \cap Y \prec Y \prec H_{\omega_3}$. We may therefore build a descending sequence $(p_n : n \in \omega)$ of conditions from $\mathbb{P}$ such that $p_0 = p$, $\{p_n : n \in \omega\} \subseteq X \cap Y$, $p_{n+1} \in D_n$ and for every $L$-cardinal $\lambda \in X \cap \gamma$ and every $\beta \in X \cap \gamma$ there is some $n \in \omega$ such that $\lambda \in suppt(p_n)$ and $\beta \in \eta_i^\lambda(p_n)$ for some $i \leq \alpha_\lambda(p_n)$. Let us write $\alpha = X \cap \omega_1$ and $q = \{(\eta_i^\lambda : i \leq \alpha) | \lambda \in X \cap \gamma$ is an $L$-cardinal$\}$ where for every $L$-cardinal $\lambda \in X \cap \gamma$, if $i < \alpha$, then $\eta_i^\lambda = \eta_i^\lambda(p_n)$ for some (all) sufficiently large $n$ and $\eta_\alpha^\lambda = X \cap \lambda$. It is not hard to check that $q \in \mathbb{P}$, $q \leq_{\mathbb{P}} p$ and $q \in D_n$ for all $n \in \omega$. $\square$

---

[3]For Harrington's club shooting forcing, I refer to Sect. 4.2.

By the above, $\mathbb{P}$ preserves $\omega_1$ and hence $\mathbb{P}$ preserves all cardinals. Let $H$ be $\mathbb{P}$-generic over $L[G]$. Now we work in $L[G, H]$. By "$(a) \Leftrightarrow (c)$" in Proposition 5.2,

$$L[G, H] \models \text{"any } \alpha \in K \text{ has the strong reflecting property"}. \qquad (5.4)$$

Thus, $K$ is a club in $\omega_2$ of $L$-cardinals with the strong reflecting property.

## 5.5   The Second Step

In this step, we work in $L[G, H]$ to find some $B_0 \subseteq \omega_2$ and $A \subseteq \omega_1$ such that $L[B_0, A] \models$ "if $\omega_1 \leq \alpha < \alpha_A$ is $A$-admissible, then $\alpha$ is an $L$-cardinal with the strong reflecting property" where $\alpha_A$ is the least $\alpha$ defined in $L[B_0, A]$ such that $L_\alpha[A] \models \mathsf{Z}_3$. Then I define a stationary set $S$ and show that $S$ contains a club.

We still work in $L[G, H]$. Note that GCH holds. Let $(B_0, \gamma^*)$ be such that

(1) $\omega_1 < \gamma^* \leq \omega_2$,
(2) $B_0 \subset \gamma^*$ and $\gamma^* = (\omega_2)^{L[B_0]}$,
(3) $L_{\gamma^*}[B_0] \prec L_{\omega_2}[G, H]$, and
(4) $\gamma^*$ is as small as possible.

Let $B$ be the theory of $(L_{\gamma^*}[B_0], B_0)$ with parameters from $\gamma^*$. i.e. $B$ denotes the subset of $\gamma^*$ coded by $T$ where $T$ is the set of pairs $(e, s)$ where $e$ is the Gödel number of a formula $\phi(x_0, \ldots, x_n)$, $s$ is a sequence $(\alpha_0, \ldots, \alpha_n)$ of ordinals $< \gamma^*$ and $\phi[\alpha_0, \ldots, \alpha_n]$ holds in $(L_{\gamma^*}[B_0], B_0)$.

*Remark 5.1*  My original definition of $B$ corresponds to the case $\gamma^* = \omega_2$, where we cannot have $\omega_2^{L[A_0]} = \omega_2^{L[B_0]}$. The definition of $(B_0, \gamma^*)$ and $B$ is choosen such that we can prove Lemma 5.3. The proof of Lemma 5.3 makes full use of our definition of $(B_0, \gamma^*)$ and $B$.

We work in $L[B_0]$. To define an almost disjoint sequence $\langle \delta_\beta^* \mid \beta < \omega_2 \rangle$ on $\omega_1$, we first define a sequence $\langle \sigma_\beta^* \mid \beta < \omega_2 \rangle$ such that for each $\beta, \sigma_\beta^*$ is the $L[B_0]$-least $\sigma \subset \omega_1$ such that $\sigma$ has cardinality $\omega_1$ and $\sigma$ is different from $\sigma_\alpha^*$ for any $\alpha < \beta$. Let $\langle s_\alpha \mid \alpha \in \omega_1 \rangle \in L[B_0]$ be an $<_{L[B_0]}$-least enumeration of $\omega_1^{<\omega_1}$. For any $\beta < \omega_2$, define $\delta_\beta^* = \{\alpha \in \omega_1 \mid \exists \eta \in \omega_1 (s_\alpha = \sigma_\beta^* \cap \eta)\}$. It is easy to check that $\langle \delta_\beta^* : \beta < \omega_2 \rangle$ is an almost disjoint sequence. By almost disjoint forcing, force $A_0 \subseteq \omega_1$ over $L[B_0]$ to code $B$ such that $\alpha \in B \Leftrightarrow |A_0 \cap \delta_\alpha^*| < \omega_1$. The forcing preserves all cardinals.

In the following, we need that $\omega_2^{L[A_0]} = \omega_2^{L[B_0]}$ which motivates Lemma 5.3.

**Lemma 5.3** (Woodin) $\omega_2^{L[A_0]} = \gamma^*$.[4]

---

[4] I would like to thank W. Hugh Woodin for pointing out the problem in our original definition of $B$ and providing this key claim.

*Proof* Let $\lambda = \omega_2^{L[A_0]}$. It follows from the definition of $(\sigma_\alpha^* : \alpha < \omega_2)$ that (i) $B_0 \cap \lambda \in L[A_0]$ and hence (ii) $\lambda = \omega_2^{L[B_0 \cap \lambda]}$. By (i) and (ii), we have (iii) $B \cap \lambda \in L[A_0]$. By the definition of $B$, it follows that (iv) $S \in L[A_0]$ where $S$ is the theory of $(L_{\gamma^*}[B_0], B_0)$ with parameters from $\lambda$. From the definition of $B$ and the fact that $A_0$ codes $B$, by (iv) it follows that $L_\lambda[B_0] \prec L_{\gamma^*}[B_0]$ and so by condition (3) in the definition of $(B_0, \gamma^*)$, we have $\lambda = \gamma^*$.  □

Next, we work in $L[A_0]$. Let $E = K \cap \{\eta \mid L_\eta[A_0] \prec L_{\omega_2}[A_0]\}$. Let

$$D = \{\gamma > \omega_1 \mid (L_\gamma[A_0, E], E \cap \gamma) \prec (L_{\omega_2}[A_0, E], E)\}.$$

Note that $D \subseteq E$. Define $F : \mathscr{P}(\omega_1) \to \mathscr{P}(\omega_1)$ as follows: if $y \subseteq \omega_1$ codes $\gamma$, then $F(y) \subseteq \omega_1$ codes $(\beta, E \cap \beta)$ where $\beta$ is the least element of $D$ such that $\beta > \gamma$ (since $D$ is a club in $\omega_2$, such $\beta$ exists); if $y$ does not code an ordinal, let $F(y) = \emptyset$.

Imitating the construction of $\langle \delta_\beta^* : \beta < \omega_2 \rangle$, we can define an almost disjoint sequence $\langle \delta_\beta \mid \beta < \omega_2 \rangle$ on $\omega_1$. We first define a sequence $\langle \sigma_\beta \mid \beta < \omega_2 \rangle$ such that for each $\beta$, $\sigma_\beta$ is the $<_{L[A_0, E]}$-least $\sigma \subset \omega_1$ such that $\sigma$ has cardinality $\omega_1$ and $\sigma$ is different from $\sigma_\alpha$ for any $\alpha < \beta$. Let $\langle t_\alpha \mid \alpha \in \omega_1 \rangle \in L[A_0, E]$ be a $<_{L[A_0, E]}$-least enumeration of $\omega_1^{<\omega_1}$. Then $\langle \delta_\beta : \beta < \omega_2 \rangle$ is a sequence of almost disjoint subset of $\omega_1$ where $\delta_\beta = \{\alpha \in \omega_1 \mid \exists \eta \in \omega_1 (t_\alpha = \sigma_\beta \cap \eta)\}$.

Let $\langle x_\alpha \mid \alpha < \omega_2 \rangle$ be the enumeration of $\mathscr{P}(\omega_1)$ in $L[A_0, E]$ in the order of construction. Define

$$Z_F = \{\alpha \cdot \omega_1 + \beta \mid \alpha < \omega_2 \wedge \beta \in F(x_\alpha)\}.$$

By almost disjoint forcing, we get $A_1 \subseteq \omega_1$ such that $\beta \in Z_F \Leftrightarrow |A_1 \cap \delta_\beta| < \omega_1$. Let $A = (A_0, A_1)$. The forcing preserves all cardinals.

Now we work in $L[B_0, A]$. Let $\alpha_A$ be the least $\alpha$ such that $L_\alpha[A] \models Z_3$. By Lemma 5.3, $\omega_1^{L[A]} = \omega_1^{L[B_0, A]}$ and $\omega_2^{L[A]} = \omega_2^{L[B_0, A]}$. Thus, $\omega_1^{L[A]} < \alpha_A < \omega_2^{L[A]}$ since $Z_3$ proves that $\omega_1$ exists. We show that in $L[B_0, A]$,

$$\text{if } \omega_1 \leq \alpha < \alpha_A \text{ is } A\text{-admissible, then } \alpha \text{ is an } L\text{-cardinal}$$

$$\text{with the strong reflecting property.} \qquad (5.5)$$

By (5.4) and Proposition 5.1, $L_{\omega_2}[G, H] \models \omega_1$ has the strong reflecting property. By condition (4) in the definition of $(B_0, \gamma^*)$ and Proposition 5.3, $L[B_0, A] \models \omega_1$ has the strong reflecting property. Assume $\omega_1 < \alpha < \alpha_A$ is $A$-admissible. Define

$$\gamma_0 = \sup(\alpha \cap D). \qquad (5.6)$$

If $\alpha \cap D = \emptyset$, let $\gamma_0 = 0$. Note that if $\gamma_0 > 0$, then $\gamma_0 \in D$. We assume that $\gamma_0 < \alpha$ and try to get a contradiction. It suffices to consider the case $\gamma_0 > 0$. Let $\alpha_0$ be the least $A_0$-admissible ordinal such that $\alpha_0 > \gamma_0$. Since $\alpha$ is $A_0$-admissible, we have $\alpha_0 \leq \alpha$.

**Lemma 5.4** $E \cap \alpha_0 = E \cap (\gamma_0 + 1)$.

*Proof* We show that $E \cap \alpha_0 \subseteq E \cap (\gamma_0 + 1)$. Suppose $\gamma \in E \cap \alpha_0$ and $\gamma > \gamma_0$. Since $\gamma \in E$, we have $L_\gamma[A_0] \prec L_{\omega_2}[A_0]$. Since $\alpha_0$ is definable from $\gamma_0$ and $A_0$, $\alpha_0$ is definable in $L_\gamma[A_0]$. Thus, $\alpha_0 \leq \gamma$ which leads to a contradiction.  $\square$

By Lemma 5.4, $L_{\alpha_0}[A_0, E] = L_{\alpha_0}[A_0, E \cap \gamma_0]$. We need the following lemma to get that $L_{\gamma_0}[A_0, E \cap \gamma_0][A_1] = L_{\gamma_0}[A]$ in Lemma 5.7.

**Lemma 5.5** $E \cap \gamma_0 \in L_{\gamma_0+1}[A]$.

*Proof* We prove by induction that for any $\gamma \in D \cap \alpha_A$, $E \cap \gamma \in L_{\gamma+1}[A]$. Fix $\gamma \in D \cap \alpha_A$. Suppose for any $\gamma' \in D \cap \gamma$, $E \cap \gamma' \in L_{\gamma'+1}[A]$. We show that $E \cap \gamma \in L_{\gamma+1}[A]$. If $\gamma \leq \omega_1$, this is trivial. Suppose $\gamma > \omega_1$.

**Case 1** There is $\gamma' \in D$ such that $\gamma$ is the least element of $D$ such that $\gamma > \gamma'$. Let $\eta$ be the least $A_0$-admissible ordinal such that $\eta > \gamma'$. By the similar argument as Lemma 5.4, $E \cap \eta = E \cap (\gamma' + 1)$. From our definitions, for any $\beta < \eta$ we have:

(1) $\langle x_\xi \mid \xi \in \beta \rangle \in L_\eta[A_0, E] = L_\eta[A_0, E \cap \gamma']$;
(2) $\langle \delta_\xi \mid \xi \in \beta \rangle \in L_\eta[A_0, E] = L_\eta[A_0, E \cap \gamma']$;
(3) $\langle x_\xi \mid \xi \in \eta \rangle$ enumerates $\mathscr{P}(\omega_1) \cap L_\eta[A_0, E] = \mathscr{P}(\omega_1) \cap L_\eta[A_0, E \cap \gamma']$.

Suppose $y \subseteq \omega_1$ and $y \in L_\eta[A_0, E \cap \gamma']$. Then $y = x_\xi$ for some $\xi < \eta$. Note that $\xi \cdot \omega + \alpha < \eta$ for any $\alpha < \omega_1$. $\alpha \in F(y)$ iff $|A_1 \cap \delta_{\xi \cdot \omega + \alpha}| < \omega_1$. Thus, $F(y) \in L_\eta[A_0, E \cap \gamma'][A_1]$. Hence we have shown that if $y \in \mathscr{P}(\omega_1) \cap L_\eta[A_0, E \cap \gamma']$, then $F(y) \in L_\eta[A, E \cap \gamma']$.

**Lemma 5.6** $L_\eta[A_0, E \cap \gamma'] \models \gamma' < \omega_2$.

*Proof* Suppose not. Then we have

$$\gamma' = \omega_2^{L_\eta[A_0, E \cap \gamma']}. \tag{5.7}$$

Let $\mathbb{P}$ be the partial order which codes $Z_F$ via $\langle \delta_\beta \mid \beta < \omega_2 \rangle$.[5] From our definitions of $E$, $F$ and $\langle x_\alpha \mid \alpha < \omega_2 \rangle$, $\mathbb{P}$ is a definable subset of $L_{\omega_2}[A_0, E]$. A standard argument gives that $\mathbb{P}$ has the $\omega_2$-c.c. in $L_{\omega_2}[A_0, E]$.[6] Let $\mathbb{P}^* = \mathbb{P} \cap L_{\gamma'}[A_0, E]$. Since $\gamma' \in D$, we have

$$(L_{\gamma'}[A_0, E], E \cap \gamma') \prec (L_{\omega_2}[A_0, E], E). \tag{5.8}$$

Suppose $D^* \subseteq \mathbb{P}^*$ is a maximal antichain with $D^* \in L_{\gamma'}[A_0, E]$. Then by (5.8), $D^*$ is a maximal antichain in $\mathbb{P}$. Since $L_{\omega_2}[A_0, E] \models |D^*| \leq \omega_1$, by (5.8), we have $L_{\gamma'}[A_0, E] \models |D^*| \leq \omega_1$. Thus, $\mathbb{P}^*$ is $\omega_2$-c.c. in $L_{\gamma'}[A_0, E]$. By (5.7), we have

$$L_\eta[A_0, E \cap \gamma'] \cap 2^{\omega_1} = L_{\gamma'}[A_0, E \cap \gamma'] \cap 2^{\omega_1}. \tag{5.9}$$

---

[5] i.e. $\mathbb{P} = [\omega_1]^{<\omega_1} \times [Z_F]^{<\omega_1}$ with $(p, q) \leq (p', q')$ iff $p \supseteq p', q \supseteq q'$ and $\forall \alpha \in q'(p \cap \delta_\alpha \subseteq p')$.
[6] i.e. If $D \subseteq \mathbb{P}$ is a maximal antichain with $D \in L_{\omega_2}[A_0, E]$, then $L_{\omega_2}[A_0, E] \models |D| \leq \omega_1$.

Since $\mathbb{P}^*$ is $\omega_2$-c.c. in $L_{\gamma'}[A_0, E]$, by (5.9), $\mathbb{P}^*$ is $\omega_2$-c.c in $L_\eta[A_0, E \cap \gamma']$.

We show that $A_1$ is generic over $L_\eta[A_0, E \cap \gamma']$ for $\mathbb{P}^*$. Let $Y \subseteq \mathbb{P}^*$ be a maximal antichain with $Y \in L_\eta[A_0, E \cap \gamma']$. Since $\mathbb{P}^*$ is $\omega_2$-c.c in $L_\eta[A_0, E \cap \gamma']$, by (5.7), $Y \in L_{\gamma'}[A_0, E \cap \gamma']$. By (5.8), $Y$ is a maximal antichain in $\mathbb{P}$. Thus, the filter given by $A_1$ meets $Y$.

Note that $\gamma' = \omega_2^{L_\eta[A_0, E \cap \gamma']} = \omega_2^{L_\eta[A_0, E \cap \gamma'][A_1]}$. Since $\gamma' \in D$, by induction hypothesis, $L_{\gamma'}[A_0, E \cap \gamma'][A_1] = L_{\gamma'}[A]$. Thus, $L_{\gamma'}[A] \models Z_3$ which contradicts the minimality of $\alpha_A$. $\square$

Take $y \in L_\eta[A_0, E \cap \gamma'] \cap \mathscr{P}(\omega_1)$ such that $y$ codes $\gamma'$. Thus, $F(y)$ codes $(\gamma, E \cap \gamma)$ and $F(y) \in L_\eta[A, E \cap \gamma']$. Then $F(y)$ is definable in $L_\gamma[A, E \cap \gamma']$. By induction hypothesis, $F(y) \in L_{\gamma+1}[A]$. Since $F(y)$ codes $E \cap \gamma$, $E \cap \gamma \in L_{\gamma+1}[A]$.

**Case 2** $\gamma$ is the least element of $D$. Take $y \in L_{\omega_1}[A_0, E] \cap \mathscr{P}(\omega_1)$ such that $y$ codes $0$. Then $y = x_0$. Since $\gamma$ is the least element of $D$ such that $\gamma > 0$, $F(y)$ codes $E \cap \gamma$. Note that for any $\beta < \omega_1$, $\langle \delta_\xi \mid \xi \in \beta \rangle \in L_{\omega_1}[A_0, E]$ and $\alpha \in F(y)$ if and only if $|A_1 \cap \delta_\alpha|$ is countable. Thus $F(y)$ is definable in $L_{\omega_1}[A, E]$. Since $E \cap \omega_1 = \emptyset$, $F(y) \in L_{\gamma+1}[A]$. Since $F(y)$ codes $E \cap \gamma$, $E \cap \gamma \in L_{\gamma+1}[A]$.

**Case 3** $\gamma$ is a limit point of $D$. Then standard argument gives that $E \cap \gamma \in L_{\gamma+1}[A]$ by induction hypothesis.

Since $\gamma_0 \in D \cap \alpha_A$, $E \cap \gamma_0 \in L_{\gamma_0+1}[A]$. $\square$

**Lemma 5.7** $L_{\alpha_0}[A_0, E \cap \gamma_0] \models \gamma_0 < \omega_2$.

*Proof* The proof is essentially the same as Lemma 5.6 (replace $\eta$ by $\alpha_0$ and $\gamma'$ by $\gamma_0$). Suppose not. Then $\gamma_0 = \omega_2^{L_{\alpha_0}[A_0, E \cap \gamma_0]}$. Let $\mathbb{P}$ be the partial order which codes $Z_F$ via $\langle \delta_\beta \mid \beta < \omega_2 \rangle$ and $\mathbb{P}^* = \mathbb{P} \cap L_{\gamma_0}[A_0, E]$. By the similar argument as Lemma 5.6, we can show that $A_1$ is generic over $L_{\alpha_0}[A_0, E \cap \gamma_0]$ for $\mathbb{P}^*$. Since $\gamma_0 = \omega_2^{L_{\alpha_0}[A_0, E \cap \gamma_0]} = \omega_2^{L_{\alpha_0}[A_0, E \cap \gamma_0][A_1]}$ and by Lemma 5.5, $L_{\gamma_0}[A_0, E \cap \gamma_0][A_1] = L_{\gamma_0}[A]$, $L_{\gamma_0}[A] \models Z_3$ which contradicts the minimality of $\alpha_A$. $\square$

From our definitions, we have

$$\text{for } \eta < \alpha_0, \langle \delta_\beta : \beta < \eta \rangle \in L_{\alpha_0}[A_0, E] = L_{\alpha_0}[A_0, E \cap \gamma_0] \text{ and} \qquad (5.10)$$

$$\langle x_\beta \mid \beta < \alpha_0 \rangle \text{ enumerates } \mathscr{P}(\omega_1) \cap L_{\alpha_0}[A_0, E] = \mathscr{P}(\omega_1) \cap L_{\alpha_0}[A_0, E \cap \gamma_0].$$
$$(5.11)$$

**Lemma 5.8** *If* $y \subseteq \omega_1$ *and* $y \in L_{\alpha_0}[A_0, E \cap \gamma_0]$, *then* $F(y) \in L_{\alpha_0}[A]$.

*Proof* Assume $y \in \mathscr{P}(\omega_1) \cap L_{\alpha_0}[A_0, E \cap \gamma_0]$. By (5.11), $y = x_\xi$ for some $\xi < \alpha_0$. Note that $\xi \cdot \omega_1 + \alpha < \alpha_0$ for $\alpha < \omega_1$. Then $\alpha \in F(y)$ iff $\xi \cdot \omega_1 + \alpha \in Z_F$ iff $|A_1 \cap \delta_{\xi \cdot \omega_1 + \alpha}| < \omega_1$. By (5.10), $F(y) \in L_{\alpha_0}[A_0, E \cap \gamma_0][A_1]$. Since by Lemma 5.5, $E \cap \gamma_0 \in L_{\gamma_0+1}[A]$, $L_{\alpha_0}[A_0, E \cap \gamma_0][A_1] = L_{\alpha_0}[A]$. Thus, $F(y) \in L_{\alpha_0}[A]$. $\square$

By Lemma 5.7, there exists $y \in L_{\alpha_0}[A_0, E \cap \gamma_0] \cap \mathscr{P}(\omega_1)$ such that $y$ codes $\gamma_0$. By the definition of $F$, $F(y)$ codes $\gamma_1$ where $\gamma_1$ is the least element of $E$ such that $\gamma_1 > \gamma_0$ and

$$(L_{\gamma_1}[A_0, E], E \cap \gamma_1) \prec (L_{\omega_2}[A_0, E], E). \tag{5.12}$$

By Lemma 5.8, $F(y) \in L_{\alpha_0}[A]$. Since $F(y)$ codes $\gamma_1$, we have $\gamma_1 < \alpha_0$. Since $\alpha_0 \leq \alpha$, we have $\gamma_1 < \alpha$. By (5.12) and (5.6), $\gamma_1 \leq \gamma_0$. Contradiction!

Thus, the assumption that $\gamma_0 < \alpha$ is false. Then $\gamma_0 = \alpha$ and hence $\alpha \in E$. By (5.4) and Proposition 5.1, we have $L_{\omega_2}[G, H] \models \alpha$ has the strong reflecting property. By condition (4) in the definition of $(B_0, \gamma^*)$ and Proposition 5.3, $L[B_0, A] \models \alpha$ has the strong reflecting property. We have proved $L[B_0, A] \models (5.5)$.

We still work in $L[B_0, A]$. Suppose $Y \prec L_{\alpha_A}[A]$, $|Y| = \omega$ and $\overline{Y}$ is the transitive collapse of $Y$. Let $\overline{\omega_1} = Y \cap \omega_1$. Then $\overline{Y} = L_{\overline{\alpha}}[\overline{A}]$ where $\overline{A} = A \cap \overline{\omega_1}$ and $\overline{\alpha} = o.t.(Y \cap \alpha_A)$. Note that $\overline{\omega_1} < \omega_1$ and $L_{\overline{\alpha}}[\overline{A}] \models Z_3$. Assume $\overline{\omega_1} \leq \eta < \overline{\alpha}$ is $\overline{A}$-admissible. By (5.5), $\eta$ is an $L$-cardinal. Let

$$Z = \{\delta < \omega_1 \mid \exists \alpha > \delta(L_\alpha[A \cap \delta] \models \text{``} Z_3 + \delta = \omega_1 + \forall \eta((\delta \leq \eta < \alpha \wedge \eta \text{ is}$$
$$A \cap \delta\text{-admissible}) \to \eta \text{ is an } L\text{-cardinal})\text{''})\}.$$

Let $Q = \{Y \cap \omega_1 \mid Y \prec L_{\alpha_A}[A] \wedge |Y| = \omega\}$. We have shown that $Q \subseteq Z$ and hence $Z$ contains a club in $\omega_1$. Define

$$S = Z \cap \{\alpha < \omega_1 : \alpha \text{ is an } L\text{-cardinal}\}. \tag{5.13}$$

Then $S$ is stationary and in fact contains a club.

## 5.6  The Third Step

In this step, we shoot a club $C$ through $S$ via Baumgartner's forcing $\mathbb{P}_S^B$ such that if $\eta$ is the limit point of $C$ and $L_{\alpha_\eta}[A \cap \eta, C \cap \eta] \models \eta = \omega_1$, then $L_{\alpha_\eta}[A \cap \eta, C \cap \eta] \models Z_3$ where $\alpha_\eta$ is the least $\alpha > \eta$ such that $L_\alpha[A \cap \eta] \models Z_3 + \eta = \omega_1$.

In this step, we use Baumgartner's forcing $\mathbb{P}_S^B$ instead of Harrington's forcing $\mathbb{P}_S$.[7] All our attempts at shooting such a club via variants of Harrington's forcing have failed: the key point is that Proposition 5.7 works for $\mathbb{P}_S^B$ but does not work for Harrington's forcing $\mathbb{P}_S$.

We still work in $L[B_0, A]$. Recall that for $f \in \mathbb{P}_S^B$ and $\eta < \omega_1$, we define in Sect. 5.3 that $(\mathbb{P}_S^B)_f = \{g \in \mathbb{P}_S^B \mid g \leq f \text{ and } \max(dom(g)) = \max(dom(f))\}$ and $\mathbb{P}_S^B \upharpoonright \eta = \{f \in \mathbb{P}_S^B \mid (dom(f) \cup ran(f)) \subseteq \eta\}$.

**Notation** For $\eta \in S$, let $\alpha_\eta$ be the least $\alpha > \eta$ such that $L_\alpha[A \cap \eta] \models Z_3 + \eta = \omega_1$.

---

[7]For details about Baumgartner's forcing $\mathbb{P}_S^B$, I refer to Sect. 5.3. For details about Harrington's forcing $\mathbb{P}_S$, I refer to Sect. 4.2.

**Lemma 5.9**

*(1) Suppose $\eta \in S$ and $\beta < \eta$. Then $L_\eta[A] \models \beta$ is countable.*

*(2) Suppose $\eta_0, \eta_1 \in S$ and $\eta_0 < \eta_1$. Then $\alpha_{\eta_0} < \eta_1$. i.e. for any $\eta \in S$, $\alpha_\eta < \bar{\eta}$ where $\bar{\eta} = \min(S \setminus (\eta + 1))$.*

*Proof* (1) Since $\eta \in S$, we have $L_{\alpha_\eta}[A \cap \eta] \models \eta = \omega_1$. Note that $\omega^\omega \cap L_{\alpha_\eta}[A \cap \eta] = \omega^\omega \cap L_\eta[A \cap \eta] = \omega^\omega \cap L_\eta[A]$. Since $\beta < \eta$, we have $L_\eta[A] = L_\eta[A \cap \eta] \models \beta$ is countable.

(2) Suppose $\eta_1 \leq \alpha_{\eta_0}$. Note that $\mathsf{Z}_3 \vdash \forall E \subseteq \omega_1 (L_{\omega_1}[E] \models \mathsf{ZFC}^-)$. Since $L_{\alpha_{\eta_1}}[A \cap \eta_1] \models \mathsf{Z}_3 + \eta_1 = \omega_1$, we have $L_{\eta_1}[A \cap \eta_0] \models \mathsf{ZFC}^-$. Since $\eta_1 \leq \alpha_{\eta_0}$ and $L_{\alpha_{\eta_0}}[A \cap \eta_0] \models \eta_0 = \omega_1$, we have $L_{\eta_1}[A \cap \eta_0] \subseteq L_{\alpha_{\eta_0}}[A \cap \eta_0]$ and hence $L_{\eta_1}[A \cap \eta_0] \models \eta_0 = \omega_1$. Since $\eta_1 \in S$, $L_{\eta_1}[A \cap \eta_0] \models \mathsf{ZFC}^-$, $L_{\eta_1}[A \cap \eta_0] \subseteq L_{\alpha_{\eta_0}}[A \cap \eta_0] \models \mathsf{Z}_3$ and $L_{\eta_1}[A \cap \eta_0] \models \eta_0 = \omega_1$, we have $L_{\eta_1}[A \cap \eta_0] \models \mathsf{Z}_3$; i.e.

$$L_{\eta_1}[A \cap \eta_0] \models \mathsf{Z}_3 + \eta_0 = \omega_1. \tag{5.14}$$

Thus, $\eta_1 \geq \alpha_{\eta_0}$ and hence $\eta_1 = \alpha_{\eta_0}$.

**Fact 5.3** (Folklore, $\mathsf{Z}_3$) *For any $E \subseteq \omega_1$, $\alpha < \omega_1$ and $a \in L_{\omega_1}[E]$, there exists $X$ such that $X \prec L_{\omega_1}[E]$, $|X| = \omega$ and $\alpha \cup \{a\} \subseteq X$.[8]*

Since $L_{\alpha_{\eta_1}}[A \cap \eta_1] \models \mathsf{Z}_3 + \eta_1 = \omega_1$, by Fact 5.3, there is $X \in L_{\alpha_{\eta_1}}[A \cap \eta_1]$ such that $X \prec L_{\eta_1}[A \cap \eta_0]$, $L_{\alpha_{\eta_1}}[A \cap \eta_1] \models |X| = \omega$, $A \cap \eta_0 \in X$ and $\eta_0 + 1 \subseteq X$ (in Fact 5.3, let $E = A \cap \eta_0, \alpha = \eta_0 + 1$ and $a = A \cap \eta_0$). Let $M$ be the transitive collapse of $X$ and $M = L_{\overline{\eta_1}}[A \cap \eta_0]$. Note that $\eta_0 < \overline{\eta_1} < \eta_1$. By (5.14), $L_{\overline{\eta_1}}[A \cap \eta_0] \models$ "$\mathsf{Z}_3 + \eta_0 = \omega_1$" and hence $\alpha_{\eta_0} \leq \overline{\eta_1} < \eta_1$, which leads to a contradiction.                                                                        $\square$

**Proposition 5.7** *Suppose $\{(\eta, \eta)\} \in \mathbb{P}_S^B$. Then $(\mathbb{P}_{S \cap \eta}^B)^{L_{\alpha_\eta}[A \cap \eta]} = \mathbb{P}_S^B \upharpoonright \eta$.*

*Proof* $\subseteq$ is trivial. Suppose $g \in \mathbb{P}_S^B \upharpoonright \eta$. We show that $g \in (\mathbb{P}_{S \cap \eta}^B)^{L_{\alpha_\eta}[A \cap \eta]}$. Let $\xi = \max(dom(g))$. Let $H : \xi + 1 \to S \cap \eta$ be the witness function for $g \in \mathbb{P}_S^B$ (i.e. $H$ is increasing, continuous and extends $g$). It suffices to find a function $H^\infty : \xi + 1 \to S \cap \eta$ such that

$$H^\infty \text{ is increasing, continuous, } H^\infty \text{ extends } g \text{ and } H^\infty \in L_{\alpha_\eta}[A \cap \eta]. \tag{5.15}$$

Pick a surjection $e_0 : \omega \to \xi + 1$ such that $e_0 \in L_{\alpha_\eta}[A \cap \eta]$ and

$$\text{for any } \alpha \leq \xi, \{i \in \omega \mid e_0(i) = \alpha\} \text{ is infinite.} \tag{5.16}$$

Pick a surjection $e_1 : \omega \to H(\xi) + 1$ such that $e_1 \in L_{\alpha_\eta}[A \cap \eta]$. Let $T$ be the set of all pairs $(\pi_1, \pi_2)$ such that $\pi_1 : k \to (H(\xi) + 1) \cap S$ where $k \in \omega$, $\pi_2 : k \to \omega$ and

---

[8]This fact is standard and its proof uses the standard Skolem Hull argument. We only need to check that the proof can be run in $\mathsf{Z}_3$ which is not hard.

the following hold[9]:

$$\text{For all } i < k, \text{ if } e_0(i) \in dom(g), \text{ then } \pi_1(i) = g(e_0(i)); \tag{5.17}$$

$$\forall i < j < k(\pi_1(i) = \pi_1(j) \Leftrightarrow e_0(i) = e_0(j)); \tag{5.18}$$

$$\forall i < j < k(\pi_1(i) < \pi_1(j) \Leftrightarrow e_0(i) < e_0(j)); \tag{5.19}$$

For all $i < k$, if $e_0(i) > 0$ is a limit ordinal and $\pi_2(i) < k$, then

$$\sup(\{e_1(m) \mid m \leq i \wedge e_1(m) < \pi_1(i)\}) < \pi_1(\pi_2(i)) < \pi_1(i) \text{ and } e_0(\pi_2(i)) < e_0(i). \tag{5.20}$$

By (5.13) and Lemma 5.9(2), we have $S \cap (H(\xi) + 1) \in L_{\alpha_\eta}[A \cap \eta]$. Since $g \in \mathbb{P}_S^B \restriction \eta$, we have $g \in L_{\alpha_\eta}[A \cap \eta]$. Since $S \cap (H(\xi) + 1), g, e_0, e_1 \in L_{\alpha_\eta}[A \cap \eta]$, by the definition of $T$, we have $T \in L_{\alpha_\eta}[A \cap \eta]$.

Define $\pi_1^\infty : \omega \to (H(\xi) + 1) \cap S$ as follows: $\pi_1^\infty(i) = H(e_0(i))$ for $i \in \omega$. Now we define $\pi_2^\infty : \omega \to \omega$ as follows such that for all $i < \omega$, if $e_0(i) > 0$ is a limit ordinal, then

$$\sup(\{e_1(m) \mid m \leq i \wedge e_1(m) < \pi_1^\infty(i)\}) < \pi_1^\infty(\pi_2^\infty(i)) < \pi_1^\infty(i) \text{ and } e_0(\pi_2^\infty(i)) < e_0(i). \tag{5.21}$$

Suppose $e_0(i) > 0$ and $e_0(i)$ is a limit ordinal. Let $\alpha = e_0(i)$. Since $H$ is continuous, $H(\alpha)$ is a limit ordinal. Let $\beta < \alpha$ be the least ordinal such that $\sup(\{e_1(m) \mid m \leq i \wedge e_1(m) < \pi_1^\infty(i)\}) < H(\beta) < H(\alpha)$. Let $\pi_2^\infty(i)$ be the least $j \in \omega$ such that $e_0(j) = \beta$. If $e_0(i) = 0$ or $e_0(i)$ is not a limit ordinal, let $\pi_2^\infty(i) = 0$. Since $\pi_1^\infty(\pi_2^\infty(i)) = \pi_1^\infty(j) = H(e_0(j)) = H(\beta)$, $\pi_1^\infty(i) = H(\alpha)$ and $e_0(\pi_2^\infty(i)) = \beta < \alpha = e_0(i)$, we have (5.21) holds.[10]

**Lemma 5.10** *For any $k \in \omega$, we have $(\pi_1^\infty \restriction k, \pi_2^\infty \restriction k) \in T$.*

*Proof* Fix $k \in \omega$. We show that $(\pi_1^\infty \restriction k, \pi_2^\infty \restriction k)$ satisfies conditions (5.17)–(5.20) in the definition of $T$. Since $H$ extends $g$, (5.17) holds. Since $H$ is strictly increasing, (5.18) and (5.19) hold. By (5.21), we have (5.20) holds. □

Define $H^\infty : \xi + 1 \to S \cap \eta$ by

$$H^\infty(e_0(i)) = \pi_1^\infty(i) \text{ for } i \in \omega. \tag{5.22}$$

We show that $H^\infty$ satisfies (5.15). By (5.18), $H^\infty$ is well-defined. By (5.19), $H^\infty$ is increasing. By (5.17), $H^\infty$ extends $g$. Since $T, e_0 \in L_{\alpha_\eta}[A \cap \eta]$, by (5.22) and Lemma 5.10, we have $H^\infty \in L_{\alpha_\eta}[A \cap \eta]$.

---

[9]The tree $T$ is defined for definability argument. We define $T$ to show that $H^\infty \in L_{\alpha_\eta}[A \cap \eta]$: we first show that $T \in L_{\alpha_\eta}[A \cap \eta]$ and then show that $H^\infty \in L_{\alpha_\eta}[A \cap \eta]$ via Lemma 5.10.

[10]To show (5.21), we use that $H$ is continuous.

**Lemma 5.11**  $H^\infty$ *is continuous.*

*Proof* Suppose $0 < \alpha \leq \xi$ is a limit ordinal. We show that $H^\infty(\alpha) = \sup(\{H^\infty(\beta) \mid \beta < \alpha\})$. Suppose not. Then there exists $\theta$ such that $\sup(\{H^\infty(\beta) \mid \beta < \alpha\}) < \theta < H^\infty(\alpha)$.

Pick $m_0$ such that $e_1(m_0) = \theta$. By (5.16), pick $i > m_0$ such that $e_0(i) = \alpha$. Since $e_1(m_0) = \theta < H^\infty(\alpha)$, by (5.22), $\theta \leq \sup(\{e_1(m) \mid m \leq i \wedge e_1(m) < \pi_1^\infty(i)\})$. By (5.22), we have $\pi_1^\infty(\pi_2^\infty(i)) = H^\infty(e_0(\pi_2^\infty(i)))$. By (5.21), we have $\theta < H^\infty(e_0(\pi_2^\infty(i)))$ and $e_0(\pi_2^\infty(i)) < e_0(i) = \alpha$. But $\sup(\{H^\infty(\beta) \mid \beta < \alpha\}) < \theta$ which leads to a contradiction.  $\square$

With this final lemma, the proof is finished.  $\square$

*Remark 5.2* The proof of Proposition 5.7 depends on (5.13) and property of Baumgartner's forcing. In fact, its proof only uses the part $(\forall \eta \in S)(\exists \delta > \eta(L_\delta[A \cap \eta] \models Z_3 + \eta = \omega_1))$ in (5.13).

**Proposition 5.8**  *Suppose* $f \in \mathbb{P}_S^B$ *where* $f = \{(\eta, \eta)\}$. *Then*

$$(\mathbb{P}_S^B)_f = \{g \cup \{(\eta, \eta)\} \mid g \in (\mathbb{P}_{S \cap \eta}^B)^{L_{\alpha_\eta}[A \cap \eta]}\}.$$

*Proof* Since $\eta \in S$, we have $\eta$ is indecomposable. This proposition follows from Lemma 5.1 and Proposition 5.7.  $\square$

Suppose $G^*$ is $\mathbb{P}_S^B$-generic over $L[B_0, A]$. Define $F_{G^*} = \bigcup\{f \mid f \in G^*\}$. Then $F_{G^*} : \omega_1 \to S$ is increasing and continuous. Let $C = ran(F_{G^*})$. Then $C \subseteq S$ is a club in $\omega_1$. Let $Lim(C) = \{\alpha \mid \alpha$ is a limit point of $C\}$. Now we work in $L[B_0, A, C]$.

**Fact 5.4**  (Folklore, [6]) *Suppose* $M \models Z_3$, $\mathbb{P} \in M$ *is a forcing notion,* $M \models$ "$|\mathbb{P}| \leq \omega_1$" *and* $G$ *is* $\mathbb{P}$-*generic over* $M$. *If* $M \models \mathbb{P}$ *preserves* $\omega_1$, *then* $M[G] \models Z_3$.

**Proposition 5.9**  *Suppose* $\eta \in Lim(C)$. *Then*

$$L_{\alpha_\eta}[A \cap \eta, C \cap \eta] \models \eta = \omega_1 \Leftrightarrow L_{\alpha_\eta}[A \cap \eta, C \cap \eta] \models Z_3.$$

*Proof* $(\Rightarrow)$: Suppose $L_{\alpha_\eta}[A \cap \eta, C \cap \eta] \models \eta = \omega_1$. Then

$$L_{\alpha_\eta}[A \cap \eta, C \cap \eta] \models C \cap \eta \text{ is a club in } \eta. \tag{5.23}$$

We show that

$$L_{\alpha_\eta}[A \cap \eta] \models S \cap \eta \text{ is stationary.} \tag{5.24}$$

Suppose not. Then there exists a club $E$ in $\eta$ such that $E \in L_{\alpha_\eta}[A \cap \eta]$ and $E \cap S \cap \eta = \emptyset$. Then $L_{\alpha_\eta}[A \cap \eta, C \cap \eta] \models E$ and $C \cap \eta$ are disjoint closed subsets of $\eta$, which leads to a contradiction.

By (5.23), $o.t.(C \cap \eta) = \eta$ and hence $\eta$ is the $\eta$-th element of $C$. Since $F_{G^*}(\xi)$ is the $\xi$-th element of $C$, $F_{G^*}(\eta) = \eta$. Let $f = \{(\eta, \eta)\}$. Since $f \in G^*$, by Proposition

5.4, $G^* \cap (\mathbb{P}^B_S)_f$ is $(\mathbb{P}^B_S)_f$-generic over $V$. By Proposition 5.8, we have $(\mathbb{P}^B_S)_f = \{h \cup \{(\eta, \eta)\} \mid h \in (\mathbb{P}^B_{S\cap\eta})^{L_{\alpha_\eta}[A\cap\eta]}\}$. Thus, $G^* \cap (\mathbb{P}^B_{S\cap\eta})^{L_{\alpha_\eta}[A\cap\eta]}$ is $(\mathbb{P}^B_{S\cap\eta})^{L_{\alpha_\eta}[A\cap\eta]}$-generic over $L_{\alpha_\eta}[A \cap \eta]$ and hence

$$C \cap \eta \text{ is } (\mathbb{P}^B_{S\cap\eta})^{L_{\alpha_\eta}[A\cap\eta]}\text{-generic over } L_{\alpha_\eta}[A \cap \eta]. \tag{5.25}$$

By (5.24), do Baumgartner's forcing $\mathbb{P}^B_{S\cap\eta}$ over $L_{\alpha_\eta}[A \cap \eta]$. Since $L_{\alpha_\eta}[A \cap \eta] \models Z_3$, by Proposition 5.5, we have $L_{\alpha_\eta}[A \cap \eta] \models$ "$|\mathbb{P}^B_{S\cap\eta}| = \omega_1$ and $\mathbb{P}^B_{S\cap\eta}$ preserves $\omega_1$". By (5.25) and Fact 5.4, $L_{\alpha_\eta}[A \cap \eta, C \cap \eta] \models Z_3$.

($\Leftarrow$): Suppose $L_{\alpha_\eta}[A \cap \eta, C \cap \eta] \models Z_3$. We show that $L_{\alpha_\eta}[A \cap \eta, C \cap \eta] \models \eta = \omega_1$. Suppose not. i.e. $\eta < \omega_1^{L_{\alpha_\eta}[A\cap\eta,C\cap\eta]}$. Since $L_{\alpha_\eta}[A \cap \eta] \subseteq L_{\alpha_\eta}[A \cap \eta, C \cap \eta]$, we have $\omega_1^{L_{\alpha_\eta}[A\cap\eta,C\cap\eta]}$ is a cardinal in $L_{\alpha_\eta}[A \cap \eta]$. But since $L_{\alpha_\eta}[A \cap \eta] \models$ "$Z_3 + \eta = \omega_1$", $\eta = \omega_1^{L_{\alpha_\eta}[A\cap\eta]}$ is the largest cardinal in $L_{\alpha_\eta}[A \cap \eta]$ which leads to a contradiction.                                                                                    □

*Remark 5.3* The key step in Proposition 5.9 is to show that (5.24) implies (5.25) which depends on the representation theorem for $(\mathbb{P}^B_{S\cap\eta})^{L_{\alpha_\eta}[A\cap\eta]}$ (Proposition 5.7).

In summary, by (5.13) and Proposition 5.9, all $\eta \in Lim(C)$ have the following properties:

$$\eta \text{ is an } L\text{-cardinal}; \tag{5.26}$$

$$\text{if } \eta \leq \beta < \alpha_\eta \text{ and } \beta \text{ is } A \cap \eta\text{-admissible, then } \beta \text{ is an } L\text{-cardinal; and} \tag{5.27}$$

$$\text{if } L_{\alpha_\eta}[A \cap \eta, C \cap \eta] \models \eta = \omega_1, \text{ then } L_{\alpha_\eta}[A \cap \eta, C \cap \eta] \models Z_3. \tag{5.28}$$

We now proceed to the fourth step.

## 5.7   The Fourth Step

In this step, I use properties of $Lim(C)$ to define the almost disjoint system on $\omega$ and some $B^* \subseteq \omega_1$. Then I do almost disjoint forcing to code $B^*$ by a real $x$. Finally, I use (5.26)–(5.28) to show that $x$ is the witness real for HP.

We still work in $L[B_0, A, C]$. Take $\alpha$ and $X$ such that $L_\alpha[A] \models Z_3$, $X \prec L_\alpha[A, C]$, $|X| = \omega$ and $X \cap \omega_1 \in Lim(C)$. Let $\eta = X \cap \omega_1$. The transitive collapse of $X$ is in the form $L_{\bar\alpha}[A \cap \eta, C \cap \eta]$. Note that $L_{\bar\alpha}[A \cap \eta] \models Z_3$ and

$$L_{\bar\alpha}[A \cap \eta, C \cap \eta] \models \eta = \omega_1. \tag{5.29}$$

By (5.29), $L_{\bar\alpha}[A \cap \eta] \models \eta = \omega_1$. Thus, $\alpha_\eta \leq \bar\alpha$. By (5.29), $L_{\alpha_\eta}[A \cap \eta, C \cap \eta] \models \eta = \omega_1$. Since $\eta \in Lim(C)$, by (5.28), $L_{\alpha_\eta}[A \cap \eta, C \cap \eta] \models Z_3$. Let $\eta^*$ be

$$\text{the least } \eta \in Lim(C) \text{ such that } L_{\alpha_\eta}[A \cap \eta, C \cap \eta] \models Z_3 + \eta = \omega_1. \tag{5.30}$$

Note that $\eta^*$ is a limit point of $Lim(C)$: Suppose not. Let $\xi < \eta^*$ be the largest element of $Lim(C)$. Then $o.t.(C \cap (\eta^* \setminus (\xi + 1))) = \omega$. But since $L_{\alpha_{\eta^*}}[A \cap \eta^*, C \cap \eta^*] \models \eta^* = \omega_1$, $L_{\alpha_{\eta^*}}[A \cap \eta^*, C \cap \eta^*] \models C \cap \eta^*$ is a club in $\eta^*$, which leads to a contradiction.

**Lemma 5.12** *Suppose* $\eta \in Lim(C)$, $\eta < \eta^*$ *and* $\beta < \alpha_\eta$. *Then* $L_{\alpha_\eta}[A \cap \eta, C \cap \eta] \models \beta < \omega_1$.

*Proof* Since $L_{\alpha_\eta}[A \cap \eta] \models Z_3$, $L_{\alpha_\eta}[A \cap \eta] \models \forall \beta \in \mathrm{Ord}(|\beta| \leq \omega_1)$. Since $L_{\alpha_\eta}[A \cap \eta] \models \eta = \omega_1$ and $\beta < \alpha_\eta$, there exists $f \in L_{\alpha_\eta}[A \cap \eta]$ such that $f : \eta \to \beta$ is surjective.

**Lemma 5.13** $L_{\alpha_\eta}[A \cap \eta, C \cap \eta] \models \eta < \omega_1$.

*Proof* Suppose $L_{\alpha_\eta}[A \cap \eta, C \cap \eta] \models \eta = \omega_1$. By (5.28), $L_{\alpha_\eta}[A \cap \eta, C \cap \eta] \models Z_3$. By (5.30), we have $\eta \geq \eta^*$, which leads to a contradiction. $\square$

Thus, there exists $g \in L_{\alpha_\eta}[A \cap \eta, C \cap \eta]$ such that $g : \omega \to \eta$ is surjective. Hence, $f \circ g : \omega \to \beta$ is surjective and $f \circ g \in L_{\alpha_\eta}[A \cap \eta, C \cap \eta]$. Thus, $L_{\alpha_\eta}[A \cap \eta, C \cap \eta] \models \beta < \omega_1$. $\square$

Now we work in $L_{\alpha_{\eta^*}}[A \cap \eta^*, C \cap \eta^*]$. We first define an almost disjoint system $\langle \delta_\beta : \beta < \eta^* \rangle$ on $\omega$ and $B^* \subseteq \eta^*$. To define $\langle \delta_\beta : \beta < \eta^* \rangle$ we first define $\langle f_\beta : \beta < \eta^* \rangle$ by induction on $\beta < \eta^*$. Let $\langle f_\beta : \omega \to 1 + \beta \mid \beta < \omega \rangle$ be an uniformly defined sequence of recursive functions.[11]

Fix $\omega \leq \beta < \eta^*$. Define $\eta_0 = \sup(Lim(C) \cap \beta)$ and $\eta_1 = \min(C \setminus (\beta + 1))$.

**Definition 5.7**

(i) Suppose $\eta_0 = 0$. Since $\eta_1 \in C$ and $\beta < \eta_1$, by Lemma 5.9(1), we have $L_{\eta_1}[A] \models \beta$ is countable. Let $f_\beta : \omega \to \beta$ be the least surjection in $L_{\eta_1}[A]$.

(ii) Suppose $\eta_0 \neq 0$ and $\beta < \alpha_{\eta_0}$. Since $\eta_0 \in Lim(C)$, $\eta_0 < \eta^*$ and $\beta < \alpha_{\eta_0}$, by Lemma 5.12, $L_{\alpha_{\eta_0}}[A \cap \eta_0, C \cap \eta_0] \models \beta < \omega_1$. Let $f_\beta : \omega \to \beta$ be the least surjection in $L_{\alpha_{\eta_0}}[A \cap \eta_0, C \cap \eta_0]$.

(iii) Suppose $\eta_0 \neq 0$ and $\beta \geq \alpha_{\eta_0}$. Since $\eta_1 \in S$ and $\beta < \eta_1$, by Lemma 5.9(1), $L_{\eta_1}[A] \models \beta$ is countable. Let $f_\beta : \omega \to \beta$ be the least surjection in $L_{\eta_1}[A]$.

Now we define an almost disjoint system $\langle \delta_\beta : \beta < \eta^* \rangle$ on $\omega$ from $\langle f_\beta : \beta < \eta^* \rangle$. Fix a recursive bijection $\pi : \omega \leftrightarrow \omega \times \omega$. Let $x_\beta = \{(i, j) \mid f_\beta(i) < f_\beta(j)\}$ and $y_\beta = \{k \in \omega \mid \pi(k) \in x_\beta\}$. Let $\langle s_i \mid i \in \omega \rangle$ be an injective, recursive enumeration of $\omega^{<\omega}$ and $\delta_\beta = \{i \in \omega \mid \exists m \in \omega(s_i = y_\beta \cap m)\}$. Then $\langle \delta_\beta : \beta < \eta^* \rangle$ is a sequence of almost disjoint reals. Since $\langle s_i \mid i \in \omega \rangle$ is recursive, $\pi$ is recursive and $f_i$ is recursive for any $i \in \omega$, we have $\langle \delta_i : i \in \omega \rangle$ is recursive.

Now we define $B^* \subseteq \eta^*$. Fix $\beta < \eta^*$. We define $z_\beta$ as follows. Let

$$\eta_0^\beta = \min(Lim(C) \setminus (\beta + 1)) \text{ and } \eta_1^\beta = \min(Lim(C) \setminus (\eta_0^\beta + 1)). \quad (5.31)$$

---

[11] Take a recursive function $F : \omega \to \omega^\omega$ such that $F(\beta)(n) = f_\beta(n)$.

Note that $\eta_1^\beta < \eta^*$ since $\eta^*$ is a limit point of $Lim(C)$. By Lemma 5.9(2), we have $\alpha_{\eta_0^\beta} < \alpha_{\eta_1^\beta}$. By Lemma 5.12, $\alpha_{\eta_0^\beta}$ is countable in $L_{\alpha_{\eta_1^\beta}}[A \cap \eta_1^\beta, C \cap \eta_1^\beta]$. Let $z_\beta$ be the least real in $L_{\alpha_{\eta_1^\beta}}[A \cap \eta_1^\beta, C \cap \eta_1^\beta]$ such that

$$z_\beta \text{ codes } \langle \eta_0^\beta, \alpha_{\eta_0^\beta}, A \cap \eta_0^\beta, C \cap \eta_0^\beta \rangle. \tag{5.32}$$

$$\text{Define } B^* = \{\omega \cdot \alpha + i \mid \alpha < \eta^* \wedge i \in z_\alpha\}. \tag{5.33}$$

By almost disjoint forcing, we get a real $x$ such that for $\alpha < \eta^*$,

$$\alpha \in B^* \Leftrightarrow |x \cap \delta_\alpha| < \omega. \tag{5.34}$$

Since $L_{\alpha_{\eta^*}}[A \cap \eta^*, C \cap \eta^*] \models Z_3$ and $x$ is a generic real built via a $c.c.c$ forcing, by Fact 5.4, we have $L_{\alpha_{\eta^*}}[A \cap \eta^*, C \cap \eta^*][x] \models Z_3$. By (5.34), (5.33) and (5.32), we have $x$ codes $(A \cap \eta^*, C \cap \eta^*)$ via $\langle \delta_\beta : \beta < \eta^* \rangle$.

We want to show that $L_{\alpha_{\eta^*}}[A \cap \eta^*, C \cap \eta^*][x] \models HP$. By absoluteness, it suffices to show in $L[B_0, A, C, x]$ that if $\lambda < \alpha_{\eta^*}$ is $x$-admissible, then $\lambda$ is an $L$-cardinal. Now we work in $L[B_0, A, C, x]$. In the rest of this section, we fix $\lambda < \alpha_{\eta^*}$ and assume that

$$\lambda \text{ is } x\text{-admissible.} \tag{5.35}$$

Since $\langle \delta_i \mid i \in \omega \rangle$ is recursive, by (5.35), we have $\langle \delta_i \mid i \in \omega \rangle \in L_\lambda[x]$. By (5.33), $B^* \cap \omega = z_0$. By (5.34), $B^* \cap \omega = \{i \in \omega \mid |x \cap \delta_i| < \omega\}$. By (5.35), $z_0 \in L_\lambda[x]$.

Define $\theta = \sup(\{\beta < \eta^* \mid z_\beta \in L_\lambda[x]\})$ and $\gamma = \sup(\{\eta_0^\beta \mid \beta < \theta\})$. By (5.31) and (5.32), for $\beta < \eta^*, z_\beta = z_{\beta+1}$. Thus, $\theta$ is a limit ordinal. By (5.32), if $\beta_0 < \beta_1 < \eta^*$, then $z_{\beta_0}$ is recursive in $z_{\beta_1}$. Thus, if $\beta < \theta$, then by (5.35), we have $z_\beta \in L_\lambda[x]$. Note that $z_\beta$ codes $(A \cap \eta_0^\beta, C \cap \eta_0^\beta)$ for $\beta < \theta$ and hence $(A \cap \gamma, C \cap \gamma) \in L_\lambda[x]$.

**Lemma 5.14** *Suppose $\theta < \lambda$. Then $\langle z_\beta \mid \beta < \theta \rangle$ is $\Sigma_1$-definable in $L_\lambda[x]$ from $(A \cap \gamma, C \cap \gamma)$.*

*Proof* If $\beta < \theta$, then $z_\beta \in L_\lambda[x]$ and hence by (5.32) and (5.35), there exists $\lambda_0 < \lambda$ such that $\lambda_0$ is a limit ordinal and $\langle \eta_0^\beta, \alpha_{\eta_0^\beta}, A \cap \eta_0^\beta, C \cap \eta_0^\beta \rangle \in L_{\lambda_0}[x]$. We can find a formula $\varphi(\alpha, z, \beta, x, A \cap \gamma, C \cap \gamma)$ which says that $\langle \eta_0^\beta, \alpha_{\eta_0^\beta}, A \cap \eta_0^\beta, C \cap \eta_0^\beta \rangle$ is countable in $L_\alpha[x]$ and $z$ is the $<_{L_\alpha[x]}$-least real which codes $\langle \eta_0^\beta, \alpha_{\eta_0^\beta}, A \cap \eta_0^\beta, C \cap \eta_0^\beta \rangle$. By absoluteness, for $\beta < \theta, z = z_\beta$ iff $\exists \lambda_0 < \lambda (z \in L_{\lambda_0}[x] \wedge \lambda_0$ is a limit ordinal $\wedge L_{\lambda_0}[x] \models \varphi[\lambda_0, z, \beta, x, A \cap \gamma, C \cap \gamma])$. $\square$

**Proposition 5.10** $\lambda$ *is an $L$-cardinal.*

*Proof* If $\beta < \theta$, then since $z_\beta$ codes $\eta_0^\beta$ and $z_\beta \in L_\lambda[x]$, by (5.35), we have $\beta < \eta_0^\beta < \lambda$. Hence $\theta \leq \lambda$ and $\gamma \leq \lambda$.

Case 1: $\theta = \lambda$. Then $\gamma = \sup(\{\eta_0^\beta \mid \beta < \lambda\}) = \lambda$. Since $\gamma \in Lim(C)$, by (5.26), we have $\lambda$ is an $L$-cardinal.

Case 2: $\theta < \lambda$. Since $(A \cap \gamma, C \cap \gamma) \in L_\lambda[x]$, by Lemma 5.14 and (5.35), we have $\langle z_\beta \mid \beta < \theta \rangle \in L_\lambda[x]$.

Subcase 1: $\alpha_\gamma \leq \lambda$. Since $\gamma, \eta^* \in Lim(C)$ and $\lambda < \alpha_{\eta^*}$, by Lemma 5.9(2), $\gamma < \eta^*$. For $i \in \omega$, since $\gamma + i < \alpha_\gamma$, by Definition 5.7(ii), $f_{\gamma+i} : \omega \to \gamma + i$ is the least surjection in $L_{\alpha_\gamma}[A \cap \gamma, C \cap \gamma]$.[12] Thus, $\langle \delta_{\gamma+i} \mid i \in \omega \rangle$ is $\Sigma_1$-definable in $L_{\alpha_\gamma}[A \cap \gamma, C \cap \gamma]$ from $(A \cap \gamma, C \cap \gamma)$. Since $\langle \delta_{\gamma+i} \mid i \in \omega \rangle$ is $\Sigma_1$-definable in $L_\lambda[x]$ from $(A \cap \gamma, C \cap \gamma)$ and $(A \cap \gamma, C \cap \gamma) \in L_\lambda[x]$, by (5.35), we have $\langle \delta_{\gamma+i} \mid i \in \omega \rangle \in L_\lambda[x]$. Note that $\omega \cdot \gamma = \gamma$ and $z_\gamma = \{i \in \omega \mid \omega \cdot \gamma + i \in B^*\} = \{i \in \omega \mid |x \cap \delta_{\gamma+i}| < \omega\}$. By (5.35), $z_\gamma \in L_\lambda[x]$ and hence $\gamma < \theta$. By the definition of $\gamma$, $\eta_0^\gamma \leq \gamma$, which leads to a contradiction.

Subcase 2: $\lambda < \alpha_\gamma$. Since $A \cap \gamma \in L_\lambda[x]$, by (5.35), $\lambda$ is $A \cap \gamma$-admissible. Since $\gamma \in Lim(C)$ and $\gamma \leq \lambda < \alpha_\gamma$, by (5.27), $\lambda$ is an $L$-cardinal. $\square$

Thus, $L_{\alpha_{\eta^*}}[A \cap \eta^*, C \cap \eta^*][x] \models Z_3 + \mathsf{HP}$ and we have proved Theorem 5.1. From [7], any remarkable cardinal is remarkable in $L$. As a corollary of Theorem 5.1, $Z_3 + \mathsf{HP}$ does not imply that $0^\sharp$ exists.

I conclude this chapter with some remarks about the proof of Theorem 5.1.

*Remark 5.4*

(1) To define an almost disjoint system on $\omega$, we usually use the reshaping technique. However, in our proof I did not use the reshaping technique and instead I use properties of $Lim(C)$ to define the almost disjoint system.

(2) If we can force a club in $\omega_2$ of $L$-cardinals with the strong reflecting property via set forcing, then we can force a set model of $Z_3 + \mathsf{HP}$ via set forcing without the use of the reshaping technique. In our proof, the hypothesis "there exists a remarkable cardinal with a weakly inaccessible cardinal above it" is only used in the first step to force a club in $\omega_2$ of $L$-cardinals with the strong reflecting property.

(3) For our proof, we need that $\omega_2$ has the strong reflecting property. Only knowing that some $\gamma \in [\omega_1, \omega_2]$ has the strong reflecting property is not enough for our proof. From this observation, only assuming one remarkable cardinal is not enough for our proof.

# References

1. Beller, A., Jensen, R.B., Welch, P.: Coding the Universe. Cambridge University Press, Cambridge (1982)
2. Cheng, Y.: Analysis of Martin-Harrington theorem in higher-order arithmetic. Ph.D. thesis, National University of Singapore (2012)
3. Baumgartner, J.: Applications of the Proper Forcing Axiom. Handbook of set-theoretic topology (Kunen, K., Vaughan, J.E.), North-Holland, Amsterdam, pp. 913–959 (1984)
4. Jech, T.J.: Set Theory. Third Millennium Edition, revised and expanded. Springer, Berlin (2003)

---

[12]This is the place we use (5.28): Definition 5.7(ii) uses Lemma 5.12 which follows from (5.28).

5. Schindler, R.: Remarkable cardinals. Infinity, Computability, and Metamathematics (Geschke et al., Eds.), Festschrift celebrating the 60th birthdays of Peter Koepke and Philip Welch, pp. 299–308
6. Woodin, W.H.: Personal communication with Woodin
7. Schindler, R.: Proper forcing and remarkable cardinals II., : Schindler, Ralf. J. Symb. Log. **66**, 1481–1492 (2001)

# Chapter 6
# The Strong Reflecting Property
# for $L$-Cardinals

**Abstract** In this chapter, I develop the full theory of the strong reflecting property for $L$-cardinals and characterize $\mathsf{SRP}^L(\omega_n)$ for $n \in \omega$ (cf. Propositions 6.7, 6.8 and Theorem 6.2). I also generalize some results on $\mathsf{SRP}^L(\gamma)$ to $\mathsf{SRP}^M(\gamma)$ for other inner models $M$ (see Theorems 6.1 and 6.4).

The notion of strong reflecting property for $L$-cardinals is introduced in Definition 5.1 and used in the proof of Theorem 5.1. However, the proof of Theorem 5.1 in Chap. 5 uses very few properties of the strong reflecting property for $L$-cardinals. In this chapter, we systematically study the strong reflecting property for $L$-cardinals. This chapter is based on some materials from [1] with some revisions and improvements.

Throughout this chapter, $\overline{\gamma}$ always denotes the image of $\gamma$ under the transitive collapse of $X$ if $X \prec H_\kappa$ and $\gamma \in X$. The distinction between $V$-cardinals and $L$-cardinals is present throughout this chapter. Whenever we write $\omega_n$ (for some $n$) without a superscript it is understood that we mean the $\omega_n$ of $V$.

**Definition 6.1** Let $\gamma \geq \omega_1$ be an $L$-cardinal.

(1) We express that $\gamma$ has the strong reflecting property for $L$-cardinals, denoted by $\mathsf{SRP}^L(\gamma)$, if for some regular cardinal $\kappa > \gamma$, if $X \prec H_\kappa$, $|X| = \omega$ and $\gamma \in X$, then $\overline{\gamma}$ is an $L$-cardinal.

(2) We express that $\gamma$ has the weak reflecting property for $L$-cardinals, denoted by $\mathsf{WRP}^L(\gamma)$, if for some regular cardinal $\kappa > \gamma$, there is $X \prec H_\kappa$ such that $|X| = \omega$, $\gamma \in X$ and $\overline{\gamma}$ is an $L$-cardinal.

If $\gamma < \omega_1$, we express that $\gamma$ has the strong reflecting property if $\gamma = \overline{\gamma}$.

**Proposition 6.1** *Suppose $\gamma \geq \omega_1$ is an $L$-cardinal. Then the following are equivalent:*

*(1)* $\mathsf{SRP}^L(\gamma)$.

*(2) For any regular cardinal $\kappa > \gamma$, if $X \prec H_\kappa$, $|X| = \omega$ and $\gamma \in X$, then $\overline{\gamma}$ is an $L$-cardinal.*

© The Author(s), under exclusive license to Springer Nature Singapore Pte Ltd. 2019    89
Y. Cheng, *Incompleteness for Higher-Order Arithmetic*, SpringerBriefs
in Mathematics, https://doi.org/10.1007/978-981-13-9949-7_6

(3) *For some regular cardinal $\kappa > \gamma$, $\{X \mid X \prec H_\kappa, |X| = \omega, \gamma \in X$ and $\bar\gamma$ is an $L$-cardinal$\}$ contains a club.*

(4) *There exists $F : \gamma^{<\omega} \to \gamma$ such that if $X \subseteq \gamma$ is countable and closed under $F$, then $o.t.(X)$ is an $L$-cardinal.*

(5) *For any regular cardinal $\kappa > \gamma$, $\{X \mid X \prec H_\kappa, |X| = \omega, \gamma \in X$ and $\bar\gamma$ is an $L$-cardinal$\}$ contains a club.*

*Proof* Note that (2) $\Rightarrow$ (1), (1) $\Rightarrow$ (3), (2) $\Rightarrow$ (5) and (5) $\Rightarrow$ (3). It suffices to show that (4) $\Rightarrow$ (2) and (3) $\Rightarrow$ (4). For the proof, I refer to Proposition 5.1.                                     $\square$

Suppose $\gamma \geq \omega_1$ is an $L$-cardinal. Let $(1)^*$, $(2)^*$, $(3)^*$, $(4)^*$ and $(5)^*$ be the statements which replace "is an $L$-cardinal" with "is not an $L$-cardinal" in Definition 6.1(1) and statements (2), (3), (4) and (5) in Proposition 6.1 respectively. The following corollary is an observation from the proof of Proposition 6.1.

**Corollary 6.1** *The statements $(1)^*$, $(2)^*$, $(3)^*$, $(4)^*$ and $(5)^*$ are equivalent.*

**Proposition 6.2** *Assume $\gamma \geq \omega_1$ is an $L$-cardinal, $\kappa$ is regular and $|\gamma| = \kappa$. Then the following are equivalent:*

(a) $\mathsf{SRP}^L(\gamma)$.

(b) *For any bijection $\pi : \kappa \to \gamma$, there exists a club $D \subseteq \kappa$ such that for any $\theta \in D$, $o.t.(\{\pi(\alpha) \mid \alpha < \theta\})$ is an $L$-cardinal.*

(c) *For some bijection $\pi : \kappa \to \gamma$, there exists a club $D \subseteq \kappa$ such that for any $\theta \in D$, $o.t.(\{\pi(\alpha) \mid \alpha < \theta\})$ is an $L$-cardinal.*

*Proof* The proof is essentially the same as the case $\kappa = \omega_1$ in Proposition 5.2.    $\square$

Let $(6)^*$ and $(7)^*$ be the statement which replaces "is an $L$-cardinal" with "is not an $L$-cardinal" in Propositions 6.2(b) and 6.2(c), respectively. The following corollary is an observation from the proof of Proposition 6.2.

**Corollary 6.2** *Suppose $\gamma \geq \omega_1$ is an $L$-cardinal, $\kappa$ is regular and $|\gamma| = \kappa$. Then the statements $(1)^*$, $(6)^*$ and $(7)^*$ are equivalent.*

**Proposition 6.3** *Suppose $\gamma \geq \omega_1$ is an $L$-cardinal. Then the following are equivalent:*

(1) $\mathsf{WRP}^L(\gamma)$.

(2) *For any regular cardinal $\kappa > \gamma$, there is $X \prec H_\kappa$ such that $|X| = \omega$, $\gamma \in X$ and $\bar\gamma$ is an $L$-cardinal.*

(3) *For some regular cardinal $\kappa > \gamma$, $\{X \mid X \prec H_\kappa, |X| = \omega, \gamma \in X$ and $\bar\gamma$ is an $L$-cardinal$\}$ is stationary.*

(4) *For any $F : \gamma^{<\omega} \to \gamma$, there exists $X \subseteq \gamma$ such that $X$ is countable, closed under $F$ and $o.t.(X)$ is an $L$-cardinal.*

(5) *For any regular cardinal $\kappa > \gamma$, $\{X \mid X \prec H_\kappa, |X| = \omega, \gamma \in X$ and $\bar\gamma$ is an $L$-cardinal$\}$ is stationary.*

*Proof* Note that (5) $\Rightarrow$ (3) and (3) $\Rightarrow$ (1). It suffices to show that (1) $\Rightarrow$ (4), (4) $\Rightarrow$ (2) and (2) $\Rightarrow$ (5). (1) $\Rightarrow$ (4) follows from (4)$^*$ $\Leftrightarrow$ (2)$^*$ in Corollary 6.1. (4) $\Rightarrow$ (2) follows from (1)$^*$ $\Leftrightarrow$ (4)$^*$ in Corollary 6.1. (2) $\Rightarrow$ (5) follows from (3)$^*$ $\Leftrightarrow$ (1)$^*$ in Corollary 6.1.                                                                                            □

**Proposition 6.4** *Suppose* $\gamma \geq \omega_1$ *is an* $L$-*cardinal,* $\kappa$ *is regular and* $|\gamma| = \kappa$. *Then the following are equivalent:*

(1) $\mathsf{WRP}^L(\gamma)$.
(2) *For some bijection* $\pi : \kappa \to \gamma$, *there exists a stationary* $D \subseteq \kappa$ *such that for any* $\theta \in D$, *o.t.*$(\{\pi(\alpha) \mid \alpha < \theta\})$ *is an* $L$-*cardinal.*
(3) *For any bijection* $\pi : \kappa \to \gamma$, *there exists a stationary* $D \subseteq \kappa$ *such that for any* $\theta \in D$, *o.t.*$(\{\pi(\alpha) \mid \alpha < \theta\})$ *is an* $L$-*cardinal.*

*Proof* Follows from Corollary 6.2 and (1)$^*$ $\Leftrightarrow$ (2)$^*$ in Corollary 6.1. The proof is standard and we omit the details.                                                                                            □

**Proposition 6.5** *Suppose* $\gamma \geq \omega_1$ *is an* $L$-*cardinal,* $\kappa > \gamma$ *is a regular cardinal and* $\mathsf{SRP}^L(\gamma)$ *holds. If* $Z \prec H_\kappa$, $|Z| \leq \omega_1$ *and* $\gamma \in Z$, *then* $\overline{\gamma}$ *is an* $L$-*cardinal.*

*Proof* Suppose $\overline{\gamma}$ is not an $L$-cardinal. Let $M$ be the transitive collapse of $Z$ and $\pi : M \prec H_\kappa$ be the inverse of the collapsing map. Take $Y \prec H_\kappa$ such that $|Y| = \omega$ and $M, \overline{\gamma} \in Y$. Note that $Y \models \overline{\gamma}$ is not an $L$-cardinal. Hence $\overline{\gamma}$ is not an $L$-cardinal.[1] Let $X = \pi``(Y \cap M)$. Since $\overline{\gamma} \in Y \cap M$ and $\pi(\overline{\gamma}) = \gamma$, we have $\gamma \in X$. Note that $X \prec Z \prec H_\kappa$ and the image of $\gamma$ under the transitive collapse of $X$ is $\overline{\gamma}$. By $\mathsf{SRP}^L(\gamma)$, $\overline{\gamma}$ is an $L$-cardinal, which leads to a contradiction.                                                         □

**Proposition 6.6** *Suppose* $\omega_1 \leq \gamma_0 < \gamma_1$ *are* $L$-*cardinals. Then* $\mathsf{SRP}^L(\gamma_1)$ *implies* $\mathsf{SRP}^L(\gamma_0)$ *(respectively* $\mathsf{WRP}^L(\gamma_1)$ *implies* $\mathsf{WRP}^L(\gamma_0)$).

*Proof* We only show the strong reflecting property case (the argument for the weak reflecting property case is similar). Let $\kappa > \gamma_1$ be a regular cardinal. It suffices to show if $X \prec H_\kappa$, $|X| = \omega$ and $\{\gamma_0, \gamma_1\} \subseteq X$, then $\overline{\gamma_0}$ is an $L$-cardinal. Note that $L_{\gamma_1} \models \gamma_0$ is a cardinal. Since $\gamma_1 \in X$, we have $L_{\gamma_1} \in X$. Since $\overline{L_{\gamma_1}} = L_{\overline{\gamma_1}}$ and $L_{\gamma_1} \models \overline{\gamma_0}$ is a cardinal, we have $L_{\overline{\gamma_1}} \models \overline{\gamma_0}$ is a cardinal. By $\mathsf{SRP}^L(\gamma_1)$, we have $\overline{\gamma_1}$ is an $L$-cardinal and hence $\overline{\gamma_0}$ is an $L$-cardinal.                                                         □

**Proposition 6.7** *The following are equivalent:*

(1) $\omega_1$ *is a limit cardinal in* $L$.
(2) $\mathsf{WRP}^L(\omega_1)$.
(3) $\mathsf{SRP}^L(\omega_1)$.

*Proof* It suffices to show that (1) $\Rightarrow$ (3) and (2) $\Rightarrow$ (1) since (3) $\Rightarrow$ (2) is immediate.

(1) $\Rightarrow$ (3): Suppose $\omega_1$ is a limit cardinal in $L$. Then $\{\alpha < \omega_1 : \alpha$ is an $L$-cardinal$\}$ is a club. By Proposition 6.2, $\mathsf{SRP}^L(\omega_1)$ holds.

(2) $\Rightarrow$ (1): Suppose $\mathsf{WRP}^L(\omega_1)$ holds. Then $\{X \cap \omega_1 : X \prec H_{\omega_2} \wedge |X| = \omega \wedge$ *o.t.*$(X \cap \omega_1)$ is an $L$-cardinal$\}$ is stationary in $\omega_1$. It is easy to see that for any $\alpha < \omega_1$ there is $\alpha < \beta < \omega_1$ such that $\beta$ is an $L$-cardinal.                                                         □

---

[1] $\overline{\overline{\gamma}}$ is the image of $\overline{\gamma}$ under the transitive collapse of $Y$.

**Proposition 6.8** *The following are equivalent:*

*(1)* $\mathsf{SRP}^L(\omega_2)$.
*(2)* $\omega_2$ *is a limit cardinal in L and for any L-cardinal* $\omega_1 \leq \gamma < \omega_2$, $\mathsf{SRP}^L(\gamma)$ *holds.*
*(3)* $\{\alpha < \omega_2 \mid \alpha$ *is an L-cardinal and* $\mathsf{SRP}^L(\alpha)$ *holds} is unbounded in* $\omega_2$.

*Proof* (1) $\Rightarrow$ (2): By Proposition 6.6, it suffices to show that $\omega_2$ is a limit cardinal in $L$. Let $\kappa > \omega_2$ be the regular cardinal that witnesses $\mathsf{SRP}^L(\omega_2)$. Fix $\alpha < \omega_2$. Pick $Z \prec H_\kappa$ such that $|Z| = \omega_1$, $\alpha \subseteq Z$ and $\omega_2 \in Z$. By Proposition 6.5, we have $\overline{\omega_2}$ is an $L$-cardinal. Note that $\alpha \leq \overline{\omega_2} < \omega_2$.

(2) $\Rightarrow$ (1): Assume $\kappa > \omega_2$ is a regular cardinal, $X \prec H_\kappa$, $|X| = \omega$ and $\omega_2 \in X$. We show that $\overline{\omega_2}$ is an $L$-cardinal. Note that $\overline{\omega_2} = o.t.(X \cap \omega_2)$. Let $E = \{\gamma \mid \omega_1 \leq \gamma < \omega_2 \wedge \gamma$ is an $L$-cardinal$\}$. $E$ is definable in $H_\kappa$. Since $\omega_2$ is a limit cardinal in $L$, we have $E$ is cofinal in $\omega_2$ and hence $E \cap X$ is cofinal in $\omega_2 \cap X$. For $\gamma \in E \cap X$, we have $\overline{\gamma} = o.t.(X \cap \gamma)$ and $\overline{\gamma}$ is an $L$-cardinal by $\mathsf{SRP}^L(\gamma)$. Note that $\overline{\omega_2} = sup(\{\overline{\gamma} \mid \gamma \in E \cap X\})$. Thus, $\overline{\omega_2}$ is an $L$-cardinal.

(1) $\Leftrightarrow$ (3): Follows from (1) $\Leftrightarrow$ (2) and Proposition 6.6.  $\square$

For definitions and properties of remarkable cardinal, I refer to Sect. 2.1.3. Any remarkable cardinal is remarkable in $L$ (cf. [2, Lemma 1.7]). Let $\kappa$ be a cardinal, $G$ be $Col(\omega, < \kappa)$-generic over $V$, $\theta > \kappa$ be a regular cardinal and $X \in [H_\theta^{V[G]}]^\omega$. Recall Definition 2.11 that $X$ condenses remarkably if $X = ran(\pi)$ for some elementary $\pi$ : $(H_\beta^{V[G \cap H_\alpha^V]}, \in, H_\beta^V, G \cap H_\alpha^V) \to (H_\theta^{V[G]}, \in, H_\theta^V, G)$ where $\alpha = crit(\pi) < \beta < \kappa$ and $\beta$ is a cardinal in $V$.

**Definition 6.2** ([2]) For regular cardinal $\theta > \kappa$, $\kappa$ is $\theta$-remarkable if in $V^{Col(\omega,<\kappa)}$, $\{X \in [H_\theta]^\omega : X$ condenses remarkably$\}$ is stationary.

From Definition 2.11, $\kappa$ is remarkable if $\kappa$ is $\theta$-remarkable for any regular cardinal $\theta > \kappa$.

**Proposition 6.9** *If $\kappa$ is remarkable in $L$ and $G$ is $Col(\omega, < \kappa)$-generic over $L$, then $L[G] \models \mathsf{WRP}^L(\gamma)$ holds for any L-cardinal $\gamma \geq \kappa$.*

*Proof* Follows from Lemma 2.2.  $\square$

Fix some $L$-cardinal $\gamma \geq \omega_1$. From Proposition 5.3, $\mathsf{SRP}^L(\gamma)$ is upward absolute (i.e. if $M \subseteq N$ are inner models and $M \models \mathsf{SRP}^L(\gamma)$, then $N \models \mathsf{SRP}^L(\gamma)$): the key point is that Proposition 6.1(4) is upward absolute. As a corollary, $\mathsf{WRP}^L(\gamma)$ is downward absolute (i.e. if $M \subseteq N$ are inner models and $N \models \mathsf{WRP}^L(\gamma)$, then $M \models \mathsf{WRP}^L(\gamma)$): the key point is that Proposition 6.3(4) is downward absolute. Thus, if $\mathsf{WRP}^L(\gamma)$ holds, then $\mathsf{WRP}^L(\gamma)$ holds in $L$. The converse is not true in general.

**Proposition 6.10** *Suppose $\mathsf{WRP}^L(\kappa)$ holds where $\kappa \geq \omega_1$ is an L-cardinal. Then $L \models$ "$\omega_1$ is $\kappa^+$-remarkable" and for any regular $\theta > \kappa$ in $L$, we have $L \models \omega_1$ is $\theta$-remarkable.*

*Proof* $L \models \mathsf{WRP}^L(\kappa)$ if and only if $\{X : X \prec L_{\kappa^+}, |X| = \omega$ and $o.t.(X \cap \kappa)$ is an $L$-cardinal$\}$ is stationary in $L$ if and only if for any $L$-regular cardinal $\theta > \kappa$, $\{X : X \prec L_\theta, |X| = \omega$ and $o.t.(X \cap \kappa)$ is an $L$-cardinal$\}$ is stationary in $L$. For $L$-regular cardinal $\theta > \kappa$, $L \models$ "$\omega_1$ is $\theta$-remarkable" if and only if for any $G$ which is $Col(\omega, < \omega_1)$-generic over $L$, we have $L[G] \models$ "$\{X \in [L_\theta]^\omega : X = ran(\pi), \pi : (L_\beta[G \restriction \alpha], \in, L_\beta, G \restriction \alpha) \prec (L_\theta[G], \in, L_\theta, G)$ where $\alpha = crit(\pi) < \beta < \omega_1$ and $\beta$ is an $L$-cardinal$\}$ is stationary". Note that $Col(\omega, < \omega_1)$ is stationary preserving and $L \models \mathsf{WRP}^L(\kappa)$. $\square$

**Corollary 6.3** *The statement "for any $L$-cardinal $\gamma \geq \omega_1$, $\mathsf{WRP}^L(\gamma)$ holds" is equiconsistent with the statement "$\omega_1$ is remarkable".*

*Proof* Follows from Propositions 6.9 and 6.10. $\square$

In the following, I generalize some results on $\mathsf{SRP}^L(\gamma)$ to $\mathsf{SRP}^M(\gamma)$ for $L$-like inner model $M$. An inner model $M$ is $L$-like if $M$ is in the form $\langle L[\vec{E}], \in, \vec{E} \rangle$ where $\vec{E}$ is a coherent sequence of extenders; moreover, for an $L$-like inner model $M$, $M|\theta$ is of the form $\langle J_\theta^{\vec{E}}, \in, \vec{E} \restriction \theta, \varnothing \rangle$.[2] In the rest of this chapter, whenever we consider an inner model $M$ we assume that $M$ is $L$-like and has the property that $M|\theta$ is definable in $H_\theta$ for any regular cardinal $\theta > \omega_2$.[3]

**Definition 6.3** Suppose $M$ is an inner model and $\gamma \geq \omega_1$ is an $M$-cardinal. We say that $\gamma$ has the *strong reflecting property* for $M$-cardinals, denoted $\mathsf{SRP}^M(\gamma)$, if for some regular cardinal $\kappa > \gamma$, if $X \prec H_\kappa, |X| = \omega$ and $\gamma \in X$, then $\bar{\gamma}$ is an $M$-cardinal.

**Definition 6.4** Suppose $M$ is an inner model. We say that $M$ has the *full covering property* if for any set $X$ of ordinals, there is $Y \in M$ such that $X \subseteq Y$ and $|Y| = |X| + \omega_1$. We say that $M$ has the rigidity property if there is no nontrivial elementary embedding from $M$ to $M$.

Recall that if $U$ is an ultrafilter on $\kappa$, $U$ is countably complete if whenever $Y \subseteq U$ is countable, we have that $\bigcap Y \neq \varnothing$.

**Theorem 6.1** *Suppose $M$ is an inner model such that $M$ has both the full covering and the rigidity property. Then, $\mathsf{SRP}^M(\gamma)$ fails for every $M$-cardinal $\gamma > \omega_2$.*

*Proof* Suppose $\mathsf{SRP}^M(\gamma)$ holds for some $\gamma > \omega_2$. Let $\kappa > \gamma$ be the witnessing regular cardinal for $\mathsf{SRP}^M(\gamma)$. Build an elementary chain $\langle Z_\alpha \mid \alpha < \omega_1 \rangle$ of submodels of $H_\kappa$ such that for all $\alpha < \beta < \omega_1$, we have $Z_\alpha \prec Z_\beta \prec H_\kappa$, $Z_\alpha \in Z_\beta$, $|Z_\alpha| = \omega$ and $\{\gamma, \omega_2\} \subseteq Z_0$.

Let $Z = \bigcup_{\alpha < \omega_1} Z_\alpha$. Then $|Z| = \omega_1$ and $Z \prec H_\kappa$. Let $\pi : N \cong Z \prec H_\kappa$ and $\pi_\alpha : N_\alpha \cong Z_\alpha \prec H_\kappa$ be the inverses of the collapsing maps. Let $j_\alpha : N_\alpha \prec N$ be the

---

[2]For the definition of coherent sequences of extenders $\vec{E}$, $J_\alpha^{\vec{E}}$ and $\vec{E} \restriction \alpha$, see [3, Sect. 2.2].
[3]All known core models satisfy this convention.

induced elementary embedding. Since $\omega_1 \subseteq Z$, we have $crit(\pi) > \overline{\omega_1}$. Since $\omega_2 \in Z$ and $|Z| = \omega_1$, we have $crit(\pi) \leq \overline{\omega_2}$. Thus, $crit(\pi) = \overline{\omega_2}$.

Note that Proposition 6.5 still holds if we replace $L$ with $M$. By $\mathsf{SRP}^M(\gamma)$, we have $\overline{\gamma}$ is an $M$-cardinal. Since $M|\overline{\gamma}$ is definable in $H_\kappa$, we have $\mathscr{P}(\overline{\omega_2}) \cap M \subseteq M|\overline{\gamma} \in N$ and $\mathscr{P}(\overline{\omega_2}) \cap M \in N$. Define $U = \{X \subseteq \overline{\omega_2} \mid X \in M \wedge \overline{\omega_2} \in \pi(X)\}$. $U$ is an $M$-ultrafilter. For $\alpha < \omega_1$, the image of $Z_\alpha$ under the transitive collapse of $Z$ is $j_\alpha``N_\alpha$ and $j_\alpha``N_\alpha \in N$. □

**Lemma 6.1** *The $M$-ultrafilter $U$ is countably complete.*

*Proof* Suppose $Y \subseteq U$ and $Y$ is countable. We show that $\bigcap Y \neq \emptyset$. Since $Y \subseteq N$, take $\alpha < \omega_1$ large enough such that $Y \subseteq j_\alpha``N_\alpha$. Let $S = \mathscr{P}(\overline{\omega_2}) \cap M \cap j_\alpha``N_\alpha$. Note that $S \in N$ and $N \models S$ is countable.

Note that $H_\kappa \models$ "$M$ has the full covering property"[4] and hence $N \models$ "$M$ has the full covering property". Fix $T \in N$ such that $T \subseteq \mathscr{P}(\overline{\omega_2}) \cap M, T \supseteq S, T \in M$ and $N \models |T| = \omega_1$. Since $\overline{\omega_2} = crit(\pi) > \omega_1$, we have $\pi(T) = \pi``T$. Since $T \in N$, we have $\mathscr{P}(T) \cap M \in N$. □

**Claim** $U \cap T \in N$.

*Proof* Since $\pi(T) = \pi``T \in M$, we have $\pi``(U \cap T) = \{\pi(A) \mid A \in T \wedge \overline{\omega_2} \in \pi(A)\} = \{B \in \pi(T) \mid \overline{\omega_2} \in B\}$ and $\pi``(U \cap T) \in M$. Note that $\mathscr{P}(\pi``T) \cap M = \pi``(\mathscr{P}(T) \cap M)$ since $\pi(D) = \pi``D$ for all $D \in \mathscr{P}(T) \cap M$. Since $\pi``(U \cap T) \in \mathscr{P}(\pi``T) \cap M$, we have $\pi``(U \cap T) = \pi(D) = \pi``D$ for some $D \in \mathscr{P}(T) \cap M \subseteq N$. Thus, $U \cap T = D$ and hence $U \cap T \in N$. □

Note that $Y \subseteq j_\alpha``N_\alpha \cap \mathscr{P}(\overline{\omega_2}) \cap M = S \subseteq T$. Since $Y \subseteq T \cap U$, to show that $\bigcap Y \neq \emptyset$, it suffices to show that $\bigcap(U \cap T) \neq \emptyset$. Note that $\overline{\omega_2} \in \bigcap \pi``(U \cap T)$ and $\pi(U \cap T) = \pi``(U \cap T)$. Then $\bigcap \pi``(U \cap T) = \bigcap \pi(U \cap T) = \pi(\bigcap(U \cap T)) \neq \emptyset$. Thus, $\bigcap(U \cap T) \neq \emptyset$. □

By the above, we can build a nontrivial embedding from $M$ to $M$ which contradicts the rigidity property of $M$. Hence, we are done. □

**Theorem 6.2** *The following are equivalent:*

(i) $\mathsf{SRP}^L(\gamma)$ *holds for some $L$-cardinal $\gamma > \omega_2$.*
(ii) $0^\sharp$ *exists.*
(iii) $\mathsf{SRP}^L(\gamma)$ *holds for every $L$-cardinal $\gamma \geq \omega_1$.*

*Proof* $(i) \Rightarrow (ii)$: Assume $0^\sharp$ does not exist. Then $L$ satisfies all the conditions for $M$ in Theorem 6.1. From the proof of Theorem 6.1 (replace $M$ with $L$), $\mathsf{SRP}^L(\gamma)$ does not hold for any $L$ cardinal $\gamma > \omega_2$.

$(ii) \Rightarrow (iii)$: Note that if $X \prec H_\kappa$ and $\gamma \in X$, then $\mathscr{M}(0^\sharp, \gamma + 1) \in X$ and its image under the transitive collapse of $X$ is $\mathscr{M}(0^\sharp, \overline{\gamma} + 1)$.[5] Note that $\mathscr{M}(0^\sharp, \alpha) \prec L$ for $\alpha \in \mathrm{Ord}$. □

---

[4]Here we use that $M|\theta$ is definable in $H_\theta$ for regular cardinal $\theta > \omega_2$.

[5]$\mathscr{M}(0^\sharp, \alpha)$ is the unique transitive $(0^\sharp, \alpha)$-model. For the definition of $\mathscr{M}(0^\sharp, \alpha)$, I refer to Sect. 2.1.2.

Thus, for $n \geq 3$, $\mathsf{SRP}^L(\omega_n)$ is equivalent to $0^{\sharp}$ exists. From Propositions 6.7, 6.8 and Theorem 6.2, we have characterized $\mathsf{SRP}^L(\omega_n)$ for $n \geq 1$.

**Definition 6.5** Suppose $M$ is an inner model. For $M$-cardinal $\lambda$, let $\mathsf{SRP}^M_{<\lambda}(\lambda)$ denote the statement: for some regular cardinal $\theta > \lambda$, if $X \prec H_\theta$, $|X| < \lambda$ and $\lambda \in X$, then $\bar{\lambda}$ is an $M$-cardinal.

**Fact 6.3** (Theorem 1.3, [4]) *Assume $0^{\dagger}$ does not exist but there is an inner model with a measurable cardinal and $L[U]$ is chosen such that $\kappa = crit(U)$ is as small as possible. The one of the following holds:*

(a) *For every set $X$ of ordinals, there is a set $Y \in L[U]$ such that $Y \supseteq X$ and $|Y| = |X| + \omega_1$;*
(b) *There is a sequence $C \subseteq \kappa$, which is Prikry generic over $L[U]$, such that for all set $X$ of ordinals, there is a set $Y \in L[U, C]$ such that $Y \supseteq X$ and $|Y| = |X| + \omega_1$.*

**Theorem 6.4** *Suppose there is an inner model with a measurable cardinal and $L[U]$ is chosen such that $\kappa = crit(U)$ is as small as possible. Suppose $\lambda > \kappa^+$ is an $L[U]$-cardinal. Then $\mathsf{SRP}^{L[U]}_{<\lambda}(\lambda)$ iff $0^{\dagger}$ exists.*

*Proof* ($\Rightarrow$) We assume that $0^{\dagger}$ does not exist and try to get a contradiction. By Fact 6.3, we need to discuss two cases.

**Cases 1** Fact 6.3.(a) holds. Let $\theta > \lambda$ be the witness regular cardinal for $\mathsf{SRP}^{L[U]}_{<\lambda}(\lambda)$. Build an elementary chain $\langle Z_\alpha \mid \alpha < \kappa \rangle$ of submodels of $H_\theta$ such that for $\alpha < \beta < \kappa$, $Z_\alpha \prec Z_\beta \prec H_\theta$, $Z_\alpha \in Z_\beta$, $|Z_\alpha| = \kappa$ and $\{\kappa^+, \lambda\} \cup tc(\{U\}) \subseteq Z_0$. Let $Z = \bigcup_{\alpha < \kappa} Z_\alpha$. Then $|Z| = \kappa$. Let $\pi : N \cong Z \prec H_\theta$ and $\pi_\alpha : N_\alpha \cong Z_\alpha \prec H_\theta$ be the inverses of the collapsing maps. Since $Z_\alpha \prec Z$, let $j_\alpha : N_\alpha \prec N$ be the induced embedding. Then $\pi_\alpha = \pi \circ j_\alpha$ and $N = \bigcup_{\alpha < \kappa} j_\alpha {}^{\prime\prime} N_\alpha$. Let $crit(\pi) = \eta$. Then $\eta > \kappa = \bar{\kappa}$ and since $|Z| = \kappa$, we have $\eta \leq \bar{\kappa}^+$. So $\eta = \kappa^+ < \bar{\lambda}$. By $\mathsf{SRP}^{L[U]}_{<\lambda}(\lambda)$, $\bar{\lambda}$ is an $L[U]$-cardinal. Let $W = \{X \subseteq \eta \mid X \in L[U]$ and $\eta \in \pi(X)\}$. Note that $U = \bar{U} \in N$ and $W \subseteq L_{\bar{\lambda}}[U] \subseteq N$. $W$ is an $L[U]$-ultrafilter on $\eta$. Note that $Z \models "|Z_\alpha| = \kappa"$ and the image of $Z_\alpha$ under the transitive collapse of $Z$ is $j_\alpha {}^{\prime\prime} N_\alpha$. Thus, for $\alpha < \kappa$, we have $j_\alpha {}^{\prime\prime} N_\alpha \in N$ and $N \models |j_\alpha {}^{\prime\prime} N_\alpha| = \kappa$.

**Lemma 6.2** *The filter $W$ is countably complete.*

*Proof* Suppose $Y \subseteq W$ and $Y$ is countable. We show that $\bigcap Y \neq \emptyset$. Since $Y \subseteq N$, take $\alpha < \kappa$ large enough such that $Y \subseteq j_\alpha {}^{\prime\prime} N_\alpha$. Let $S = \mathscr{P}(\eta) \cap L[U] \cap j_\alpha {}^{\prime\prime} N_\alpha$. Note that $\mathscr{P}(\eta) \cap L[U] \in N$ and hence $S \in N$. We have $N \models |S| \leq \kappa$. Since Fact 6.3(a) holds in $H_\theta$ and $N \prec H_\theta$, we have Fact 6.3(a) holds in $N$. Take $T \in N$ such that $T \subseteq \mathscr{P}(\eta) \cap L[U]$, $T \supseteq S$, $T \in L[U]$ and $N \models |T| \leq \kappa$. Since $\eta > \kappa$, we have $\pi(T) = \pi {}^{\prime\prime} T$. Let $\bar{T} = \{X \in T \mid \eta \in \pi(X)\}$.

**Claim** $\bar{T} \in N$.

*Proof* Since $N \models |T| \leq \kappa$, there is $h \in N$ such that $h : T \leftrightarrow \gamma$ for some $\gamma < \eta$. Then $\bar{T} = \{X \in T \mid \eta \in \pi {}^{\prime\prime}(h^{-1})(h(X))\}$. Thus, $\bar{T} \in N$. □

Note that $\bigcap \overline{T} \neq \emptyset$ since $\pi(\overline{T}) = \pi^{\omega}\overline{T}$ and $\eta \in \bigcap \pi^{\omega}\overline{T} = \bigcap \pi(\overline{T}) = \pi(\bigcap \overline{T})$. Since $Y \subseteq S \subseteq T$ and $Y \subseteq W$, we have $Y \subseteq \overline{T}$ and hence $\bigcap Y \neq \emptyset$. $\qquad \square$

Thus, there exists a nontrivial elementary embedding $j : L[U] \prec L[U]$ with $crit(j) = \eta > \kappa$. By Fact 2.16, $0^{\dagger}$ exists which leads to a contradiction.

**Cases 2** Fact 6.3(b) holds. The proof is essentially the same as Case 1 with small modifications (let $tc(\{U, C\}) \subseteq Z_0$ and $W = \{X \subseteq \eta \mid X \in L[U, C]$ and $\eta \in \pi(X)\}$). Since Priky forcing preserves all cardinals, we have $\overline{\lambda}$ is an $L[U, C]$-cardinal. As in Case 1, we can show that there exists a nontrivial elementary embedding $j : L[U, C] \prec L[U, C]$. Since $j(U, C) = (U, C)$, we have $j \upharpoonright L[U] : L[U] \prec L[U]$. Note that $crit(j \upharpoonright L[U]) = \eta > \kappa$. Thus, by Fact 2.16, $0^{\dagger}$ exists which leads to a contradiction.

($\Leftarrow$) Assume $0^{\dagger}$ exists. Suppose $\theta > \lambda$ is regular, $X \prec H_{\theta}$, $|X| < \lambda$ and $\lambda \in X$. We show that $\overline{\lambda}$ is an $L[U]$-cardinal. Since $\lambda \in X$ and $0^{\dagger} \in X$, we have $\mathscr{M}(0^{\dagger}, \omega, \lambda + 1) \in X$.[6] Note that for any $\alpha, \beta \in \mathbf{Ord}$, we have $\mathscr{M}(0^{\dagger}, \alpha, \beta) \prec L[U]$. Since $\lambda$ is an $L[U]$-cardinal and $\lambda \in \mathscr{M}(0^{\dagger}, \omega, \lambda + 1)$, we have $\mathscr{M}(0^{\dagger}, \omega, \lambda + 1) \models \lambda$ is a cardinal. Note that the image of $\mathscr{M}(0^{\dagger}, \omega, \lambda + 1)$ under the transitive collapse of $X$ is $\mathscr{M}(0^{\dagger}, \omega, \overline{\lambda} + 1)$. Thus, $\mathscr{M}(0^{\dagger}, \omega, \overline{\lambda} + 1) \models \overline{\lambda}$ is a cardinal. Since $\mathscr{M}(0^{\dagger}, \omega, \overline{\lambda} + 1) \prec L[U]$, we have $\overline{\lambda}$ is an $L[U]$-cardinal. $\qquad \square$

There are principles in the literature that are similar (or even equivalent) to the strong reflecting property for $L$-cardinals from Sect. 5.2. In the rest of this chapter, I give two examples of such principles and examine the relationship between them and the notion of strong reflecting property for $L$-cardinals.

In [6], Schindler introduces the principle $S_{\lambda}(L)$ for $L$-cardinal $\lambda > \omega_1$ and define $S_{\lambda}(L) = \{X \prec L_{\lambda} : |X| = \omega \wedge X \cap \omega_1 \in \omega_1 \wedge X \cong L_{\theta}$ for some $L$-cardinal $\theta\}$. Schindler [6] proved the following facts about $S_{\lambda}(L)$:

**Theorem 6.5** *(1) (Lemma 2.3, [6]) The set $S_{\omega_2}(L)$ contains a club if and only if [$\omega_2$ is inaccessible in $L$ and $S_{\lambda}(L)$ contains a club for every $L$-cardinal $\lambda$ such that $\omega_1 < \lambda < \omega_2$].*
*(2) (Lemma 2.2, [6]) Let $\lambda > \omega_2$ be an $L$-cardinal. Then $S_{\lambda}(L)$ contains a club if and only if $0^{\sharp}$ exists.*

Note that Theorem 6.5 is a reformulation of Proposition 6.8 and Theorem 6.2.

In [7], Räsch and Schindler introduced the condensation principle $\nabla_{\kappa}$: for any regular cardinal $\theta > \kappa$, $\{X \prec L_{\theta} \mid |X| < \kappa, X \cap \kappa \in \kappa$ and $L \models o.t.(X \cap \theta)$ is a cardinal$\}$ is stationary. The notion of the strong reflecting property for $L$-cardinals was introduced before the author knew about the work on $\nabla_{\kappa}$ in [7]. The following theorem summarizes the strength of $\nabla_{\omega_n}$ for $n \in \omega$.

**Theorem 6.6** *(1) (Theorem 2, 4, [7]) The following are equiconsistent:*

*(a)* $\mathsf{ZFC} + \nabla_{\omega_1}$.

---

[6]Note that $\mathscr{M}(0^{\dagger}, \omega, \alpha)$ is the unique transitive $(0^{\dagger}, \omega, \alpha)$-model. For the definition of $\mathscr{M}(0^{\dagger}, \omega, \alpha)$, I refer to [5].

*(b)* ZFC $+ \nabla_{\omega_2}$.
*(c)* ZFC $+$ *there exists a remarkable cardinal.*

*(2)  (Corollary 12, [7]) For $n \geq 3$, $\nabla_{\omega_n}$ is equivalent to $0^\sharp$ exists.*

I now discuss the relationship between $\mathsf{SRP}^L(\omega_n)$ and $\nabla_{\omega_n}$ for $n \in \omega$. By Theorem 6.2 and Theorem 6.6, $\mathsf{SRP}^L(\omega_n)$ is equivalent to $\nabla_{\omega_n}$ for $n \geq 3$. If $\kappa$ is regular cardinal and $\nabla_\kappa$ holds, then $\kappa$ is remarkable in $L$ (cf. [7, Lemma 7]). By Proposition 6.7, $\nabla_{\omega_1}$ implies $\mathsf{SRP}^L(\omega_1)$ which is strictly weaker. In Definition 6.1, we only consider countable elementary submodels of $H_\kappa$. Similarly as $\nabla_\kappa$ we could also consider uncountable elementary submodels of $H_\kappa$. However this does not change the picture. Note that $\mathsf{SRP}^L_{<\omega_1}(\omega_1)$ iff $\mathsf{SRP}^L(\omega_1)$. By Proposition 6.5, we have $\mathsf{SRP}^L_{<\omega_2}(\omega_2)$ iff $\mathsf{SRP}^L(\omega_2)$. By Theorem 6.1, for $n \geq 3$, $\mathsf{SRP}^L_{<\omega_n}(\omega_n) \Leftrightarrow 0^\sharp$ exists $\Leftrightarrow \mathsf{SRP}^L(\omega_n)$.

# References

1. Cheng, Y.: The strong reflecting property and Harrington's Principle. Math Logic Quart. **61**(4–5), 329–340 (2015)
2. Schindler, Ralf: Proper forcing and remarkable cardinals II. J. Symb. Log. **66**, 1481–1492 (2001)
3. Steel, R.J.: An outline of inner model theory. In: Foreman, M., Kanamori, A. (Eds.) Chapter 19 in Handbook of Set Theory. Springer, Berlin (2010)
4. Mitchell, W.J.: The covering lemma. In: Foreman, M., Kanamori, A. (Eds.) Chapter 18 in Handbook of Set Theory. Springer, Berlin (2010)
5. Kanamori, A.: Higher Infinite: Large Cardinals in Set Theory from Their Beginnings, 2nd Edn. Springer Monographs in Mathematics, Springer, Berlin (2003)
6. Schindler, R.: Remarkable cardinals. Infinity, Computability, and Metamathematics (Geschke et al., eds.), Festschrift celebrating the 60th birthdays of Peter Koepke and Philip Welch, pp. 299–308
7. Räsch, T., Schindler, R.: A new condensation principle. Archive Math. Logic. **44**, 159–166 (2005)

# Appendix A
# Sami's Forcing-Free Proof in $Z_2$

**Abstract** In this appendix, I reconstruct in $Z_2$ Sami's forcing-free proof of "$Det(Turing\text{-}\Sigma_1^1)$ implies HP" in a general setting.

In Sect. 1.2, I give a brief survey of different proofs of "$Z_2 + Det(\Sigma_1^1)$ implies HP" in the literature. All other proofs of Harrington's Theorem in the literature use forcing. Sami's proof of Harrington's Theorem in [1] is totally forcing-free and only uses effective descriptive set theory. All definitions and results in this section are due to Sami (cf. [1]), only the reconstruction in this form is due to the author.

The structure of this appendix is as follows. I first introduce a general property, called '$\bigstar$', and then show in Theorem A.3 that if there is a $\Sigma_1^1$ set of reals with property $\bigstar$, then $Z_2 + Det(Turing\text{-}\Sigma_1^1)$ implies HP. Then I present Sami's definition of a $\Sigma_1^1$ set of reals $D$ and show that $D$ has the $\bigstar$ property. Hence, by Theorem A.11, we have $Z_2 + Det(Turing\text{-}\Sigma_1^1)$ implies HP.

**Definition A.1** A set $Z \subseteq \omega^\omega$ is said to have '*property $\bigstar$*' if

(1) $Z$ is Turing closed and cofinal in the Turing degree;
(2) If $C$ is a Turing cone with $C \subseteq Z, x \in C, \beta < \omega_1^x$ and $y \subseteq \beta$, then $y \in L$ implies $y \in L_{\omega_1^x}[x]$.

Now, $A \subseteq \omega^\omega$ is cofinal in the Turing degrees if $\forall x \in \omega^\omega \exists y \in A(x \leq_T y)$. Given a real $x$, the Turing cone of $x$ is $Cone(x) = \{y \in \omega^\omega \mid x \leq_T y\}$. We say $A \subseteq \omega^\omega$ contains a cone if for some real $x$, $Cone(x) \subseteq A$.

**Lemma A.1** (Martin's Lemma) *For a Turing-closed set $A \subseteq \omega^\omega$, $A$ is determined iff $A$ or its complement includes a cone.*

*Proof* The converse direction is easy. We only prove the forward direction. Suppose $A \subseteq \omega^\omega$ is Turing-closed and $A$ is determined, there is a real $x$ such that $Cone(x) \subseteq A$ or $Cone(x) \cap A = \emptyset$. We show that if player I has a winning strategy in the game $G_A$, then $A$ contains a cone. Let $\sigma$ be a winning strategy for player I. It suffices to show that $A$ contains the cone $\{x \in \omega^\omega : \sigma \leq_T x\}$. Let $x \in \omega^\omega$ be such that $\sigma \leq_T x$.

Then $a = \sigma * x$ is in $A$ because $\sigma$ is a winning strategy for player I. Since $x \equiv_T a$ and $A$ is $\equiv_T$-closed, we have $x \in A$. Similarly, we can show that if player II has a winning strategy in the game $G_A$, then the complement of $A$ contains a cone. □

By Lemma A.1, $Det(Turing-\Gamma)$ is the statement that if $X \subseteq \omega^\omega$ is in $\Gamma$ and Turing closed, then:

$$[\exists x \forall y(x \leq_T y \to y \in X)] \vee [\exists x \forall y(x \leq_T y \to y \notin X)].$$

**Theorem A.2** (G. Sacks, [2]) *If $x \in \omega^\omega$ and $\alpha$ is a countable $x$-admissible ordinal, then $\alpha = \omega_1^y$ for some $y \in Cone(x)$.*

**Theorem A.3** ($Z_2$) *Suppose there is a $\Sigma_1^1$ set of reals with property ★. Then $Det(Turing-\Sigma_1^1)$ implies HP.*

*Proof* Assume that $Z$ is a $\Sigma_1^1$ set of reals with property ★ and $Det(Turing-\Sigma_1^1)$ holds. Since $Z$ is Turing closed and cofinal in the Turing degree, by Lemma A.1, $Z$ contains a cone of Turing degrees. Suppose $C = Cone(c) \subseteq Z$. Now we show that for any countable ordinal $\alpha$, if $\alpha$ is $c$-admissible, then $\alpha$ is an $L$-cardinal.

Let $\alpha$ be a countable $c$-admissible ordinal. By Sacks' Theorem (cf. Theorem A.2), $\alpha = \omega_1^{c \oplus y}$ for some $y \in \omega^\omega$. Thus, $\alpha = \omega_1^y$ for some $y \in C$.

**Claim** *If $z \in C$, then $\omega_1^z$ is an $L$-cardinal.*

*Proof* If $\omega_1^z$ is not an $L$-cardinal, then for some $\alpha < \omega_1^z$, there is a well-ordering $R \subseteq \alpha \times \alpha$ such that $R \in L$ and $o.t.(R) = \omega_1^z$. Let $\gamma = \alpha \times \alpha$. Then $\gamma < \omega_1^z$ and $R \subseteq \gamma$. Since $Z$ has property ★, we have $R \in L_{\omega_1^z}[z]$. Since $o.t.(R) = \omega_1^z$, by admissibility, we have $\omega_1^z \in L_{\omega_1^z}[z]$, which leads to a contradiction. □

Since $y \in C$, $\alpha = \omega_1^y$ is an $L$-cardinal. □

In the following, we define such a $\Sigma_1^1$ set of reals with property ★. First of all, I review some definitions and facts we will use in the following. $\mathcal{R} = \mathcal{P}(\omega \times \omega)$ is the space of relations on $\omega$ and $S_\infty$ is the space of permutations of $\omega$, each equipped with its usual recursively presented Polish topology. Let $S \subseteq \mathcal{X} \times \mathcal{Y}$ where $\mathcal{Y}$ is a topological space. The category quantifier $\exists^* y(S(x, y))$ stands for: $\{y \in \mathcal{Y} \mid S(x, y)\}$ is non-meager in $\mathcal{Y}$.

Linear orderings are taken to be reflective, i.e. non-strict. $LO = \{r \subseteq \omega \times \omega \mid r$ is a linear ordering on its field$\}$. For $r \in LO$, $\leq_r$ is just binary relation $r$ and $<_r$ has its usual meaning. Note that $WO = \{r \in LO :<_r$ is well-founded$\}$, and $WO_\alpha = \{r \in WO \mid |r| = \alpha\}$ for $\alpha < \omega_1$. For $r \in \mathcal{R}$ and $k \in \omega$, $r \upharpoonright k = \{(m, n) \mid m <_r k \wedge n <_r k \wedge m \leq_r n\}$. Note that $r \upharpoonright k = \emptyset$ if $k \notin Field(r)$, and the function $(r, k) \mapsto r \upharpoonright k$ is recursive.

Given $f \in S_\infty$ and $r \subseteq \omega \times \omega$, we denote by $f \cdot r$ the isomorphic copy of $r$ by $f$. Note that $(f, r) \mapsto f \cdot r$ is a recursive function from $S_\infty \times \mathcal{R}$ to $\mathcal{R}$. Suppose $r, s \subseteq \omega \times \omega$ are isomorphic via $g : (\omega, r) \to (\omega, s)$. For any $Z \subseteq \mathcal{R}$, we have $\{f : f \cdot r \in Z\} = \{f : f \cdot s \in Z\} \circ g$ since $(f \circ g^{-1}) \cdot s \in Z$ if $f \cdot r \in Z$, and $(f \circ g) \cdot$

$r \in Z$ if $f \cdot s \in Z$.[1] Since right multiplication by $g$ is a homeomorphism of $S_\infty$, the topological properties of $\{f : f \cdot r \in Z\}$ and $\{f : f \cdot s \in Z\}$ are identical.

**Definition A.4** ([1]) $r \in$ LO is a pseudo-well-ordering if any non-empty $\Delta_1^1(r)$ subset of Field($r$) has a $r$-least element. Let PWO denote the set of such orderings.

Note that WO $\subseteq$ PWO and PWO is $\Sigma_1^1$.

**Fact A.5** (Harrison, [3]) If $r \in$ PWO\WO, then $o.t.(r) = \omega_1^r \times (1 + \eta) + \rho_r$ where $\eta$ is the order type of the rationals and $\rho_r < \omega_1^r$.

**Fact A.6** (Boundedness theorem for $\Sigma_1^1$ ($\Sigma_1^1$) set, [2, 4])

(1) If $A \subseteq$ WO is $\Sigma_1^1$, then there is an $\alpha < \omega_1$ such that $A \subseteq WO_{<\alpha}$;
(2) If $A \subseteq$ WO is $\Sigma_1^1$, then there is an $\alpha < \omega_1^{CK}$ such that $A \subseteq WO_{<\alpha}$. As a corollary, if $A \subseteq$ WO and $A$ is $\Sigma_1^1$, then $\sup\{rk(x) : x \in A\} < \omega_1^{CK}$.

**Proposition A.1** (Sami, [1]) If $r \in$ PWO and $\omega_1^r = \omega_1^{CK}$, then $r$ has an isomorphic recursive copy.

*Proof* If $r \in$ WO, then $rk(r) < \omega_1^{CK}$. Thus, $r$ has an isomorphic recursive copy. If $r \in$ PWO\WO, by Fact A.5, we have $o.t.(r) = \omega_1^{CK} \times (1 + \eta) + \rho_r$ where $\eta$ is the order type of the rationals and $\rho_r < \omega_1^{CK}$.

From Fact A.6, $\{r \in$ WO $: r$ is recursive$\}$ is not $\Sigma_1^1$ (if not, then $\sup\{rk(x) : x$ is a recursive well-ordering on $\omega\} < \omega_1^{CK}$ which leads to a contradiction). But $\{r \in$ PWO $: r$ is recursive $\}$ is $\Sigma_1^1$. Take a recursive $s \in$ PWO\WO. Since $s$ is recursive, by trimming some excess, we can assume that $o.t.(s) = \omega_1^{CK} \times (1 + \eta)$ where $\eta$ is the order type of the rationals. Thus, by stringing together $s$ and a recursive well-ordering with length $\rho_r$, we can construct a recursive copy of $r$. $\square$

**Fact A.7** (Kechris, [4]) If $R \subseteq \mathcal{X} \times \mathcal{Y}$ is $\Sigma_\alpha^0$ with $\alpha < \omega_1^{CK}$(resp. $R$ is $\Delta_1^1$) where $\mathcal{X}, \mathcal{Y}$ are recursively presented Polish spaces, then the relation $\exists^* y(R(-, y))$ is $\Sigma_\alpha^0$ (resp. $\Delta_1^1$).

**Fact A.8** ([5, 6]) For each $n \geq 1$, there exists a universal $\Sigma_n^1(\Sigma_n^1)$ set in $\mathcal{N}^2$; i.e. a set $U \subseteq \mathcal{N}^2$ such that $U$ is $\Sigma_n^1(\Sigma_n^1)$ and that for every $\Sigma_n^1(\Sigma_n^1)$ set $A \subseteq \mathcal{N}$ there exists some $y \in \mathcal{N}$ such that $A = \{x : (x, y) \in U\}$.

Given Polish space $X$, $A \subseteq X$ is nowhere dense if the complement of $A$ contains a dense open set. $A \subseteq X$ is meager (or of first category) if $A$ is the union of countably many nowhere dense sets. A non-meager set is called a set of second category.

**Fact A.9** (Baire category theorem, [6]) In a Polish space, every nonempty open set is non-meager (or of second category).

**Fact A.10** (Gandy's basis theorem, [2, 7]) Let $A \subseteq \omega^\omega$ and $z \in \omega^\omega$.

---

[1] $f \circ g$ means the composition of $f$ and $g$.

*(1) If A is non-empty $\Sigma_1^1(z)$ set, then there is $x \in A$ such that $\omega_1^{x \oplus z} = \omega_1^z$.*
*(2) If A is non-empty $\Sigma_1^1$ set, then there exists $a \in A$ such that $\omega_1^a = \omega_1^{\mathsf{CK}}$.*

**Theorem A.11** (Sami, [1]) *Given $a \in \omega^\omega$ and $1 \leq \alpha < \omega_1^{\mathsf{CK}}$, if for some $\eta < \omega_1$, $a$ is $\Sigma_\alpha^0(x)$ for all $x \in \mathsf{WO}_\eta$, then $a$ is $\Sigma_\alpha^0$.*

*Proof* Let $U \subseteq \omega \times \mathscr{R} \times \omega$ be $\omega$-universal for $\Sigma_\alpha^0$ subsets of $\mathscr{R} \times \omega$. Fix $r \in \mathsf{WO}_\eta$. By the hypothesis, for all $f \in S_\infty$ there is $e \in \omega$ such that $a = U(e, f \cdot r, -)$ (i.e. $n \in a \Leftrightarrow U(e, f \cdot r, n)$).

Since $S_\infty = \{f \in S_\infty \mid \exists e \in \omega (a = U(e, f \cdot r, -))\} = \bigcup_{n \in \omega} \{f \in S_\infty \mid a = U(n, f \cdot r, -)\}$, by Baire category theorem (cf. Fact A.9), there exists an $e_0 \in \omega$ such that $\{f \in S_\infty \mid a = U(e_0, f \cdot r, -)\}$ is non-meager in $S_\infty$.

Suppose toward a contradiction that $a$ is not $\Sigma_\alpha^0$. Let

$$A = \{(x, s) \mid x \in \omega^\omega \text{ is not } \Sigma_\alpha^0, s \in \mathsf{PWO} \text{ and } \exists^* f \in S_\infty(x = U(e_0, f \cdot s, -))\}. \tag{A.1}$$

Note that "$x$ is $\Sigma_\alpha^0$" is a $\Delta_1^1$ property of $x$ and "$x = U(e_0, f \cdot s, -)$" is a $\Delta_1^1$ property of $(f, x, s)$. By Fact A.7, we have "$\exists^* f \in S_\infty(x = U(e_0, f \cdot s, -))$" is $\Delta_1^1$ and hence $A$ is $\Sigma_1^1$. $A$ is non-empty since $(a, r) \in A$. By Gandy's basis theorem (cf. Fact A.10), pick $(x_0, s_0) \in A$ such that $\omega_1^{(x_0, s_0)} = \omega_1^{\mathsf{CK}}$. Thus, $\omega_1^{s_0} = \omega_1^{\mathsf{CK}}$.

Since $s_0 \in \mathsf{PWO}$, by Proposition A.1, let $w_0$ be a recursive copy of $s_0$. Since $\{f \mid x_0 = U(e_0, f \cdot w_0, -)\}$ is a translate in $S_\infty$ of $\{f \mid x_0 = U(e_0, f \cdot s_0, -)\}$, we have $\exists^* f \in S_\infty(x_0 = U(e_0, f \cdot w_0, -))$. Let $V \subseteq S_\infty$ be a non-empty basic open set such that $\{f \mid x_0 = U(e_0, f \cdot w_0, -)\}$ is co-meager in $V$.

**Claim**

$$n \in x_0 \Leftrightarrow (\exists^* f \in V) U(e_0, f \cdot w_0, n).$$

*Proof* Define $P = \{f \in V \mid U(e_0, f \cdot w_0, n)\}$, $Q = \{f \mid x_0 = U(e_0, f \cdot w_0, -)\}$ and $T = \{f \in V \mid x_0 = U(e_0, f \cdot w_0, -)\}$.

($\Rightarrow$): Since $V \backslash Q$ is meager, $T$ is non-meager. Since $n \in x_0$, we have $T \subseteq P$. Thus, $P$ is non-meager.

($\Leftarrow$): If $\exists^* f \in V(U(e_0, f \cdot w_0, n))$, then $P$ is non-meager. Suppose $Q \cap P = \emptyset$. Then $P \subseteq V \backslash Q$. Since $Q$ is co-meager in $V$ if and only if $V \backslash Q$ is meager, we have $P$ is meager which leads to a contradiction. Thus $Q \cap P \neq \emptyset$ and $n \in x_0$. □

Since $w_0$ is recursive, "$U(e_0, f \cdot w_0, n)$" as a relation in $(f, n)$ is $\Sigma_\alpha^0$. Thus "$\exists^* f \in V(U(e_0, f \cdot w_0, n))$" is $\Sigma_\alpha^0$. Hence $x_0$ is $\Sigma_\alpha^0$. This contradicts the definition of $A$ in (A.1), and we are done. □

*Remark A.1* The proof of Theorem A.11 in fact shows that if $\{f \mid a \text{ is } \Sigma_\alpha^0(f \cdot r)\}$ is non-meager in $S_\infty$ for some $r \in \mathsf{PWO}$, then $a$ is $\Sigma_\alpha^0$.

**Proposition A.2** (Sami, [1]) *Suppose $\alpha < \omega_1$. Then:*

*(1) $\mathsf{WO}_\alpha$ is $\Sigma_{\alpha+2}^0$;*
*(2) Given $r \in \mathsf{WO}_\alpha$, the relation "$s \in \mathsf{WO}_{|r \restriction k|}$" in $(s, k)$ is $\Sigma_{\alpha+2}^0(r)$.*

*Proof* Prove (1) by induction on $\alpha$. If $\alpha$ is a limit ordinal, then

$$r \in \mathsf{WO}_\alpha \Leftrightarrow (\bigwedge_{\beta < \alpha} \bigvee_{n \in \omega} (r \upharpoonright n \in \mathsf{WO}_\beta)) \wedge (\bigwedge_{n \in \omega} \bigvee_{\beta < \alpha} (r \upharpoonright n \in \mathsf{WO}_\beta)).$$

By induction, "$r \upharpoonright n \in \mathsf{WO}_\beta$" is $\Sigma^0_{\beta+2}$. Thus, "$\bigvee_{\beta < \alpha} (r \upharpoonright n \in \mathsf{WO}_\beta)$" is $\Sigma^0_\alpha$. Similarly, "$\bigwedge_{\beta < \alpha} \bigvee_{n \in \omega} (r \upharpoonright n \in \mathsf{WO}_\beta)$" is $\Pi^0_\alpha$. Hence $\mathsf{WO}_\alpha$ is $\Sigma^0_{\alpha+2}$.
If $\alpha = \beta + 1$, then

$$r \in \mathsf{WO}_\alpha \Leftrightarrow r \in \mathsf{LO} \wedge \exists n(n \text{ is } \leq_r \text{-maximum} \wedge r \upharpoonright n \in \mathsf{WO}_\beta). \tag{A.2}$$

The R.H.S of (A.2) is $\Sigma^0_{\beta+2}$ and hence $\mathsf{WO}_\alpha$ is $\Sigma^0_{\alpha+2}$. Finally, (2) is just the effective version of (1), and we are done. $\qquad \square$

**Fact A.12** ([1]) *Given* $r, s \in \mathsf{LO}$ *with the same order type, there is* $s' \leq_T s$ *such that* $(\omega, r) \cong (\omega, s')$.

Given $\alpha < \omega_1, r \in \mathsf{WO}_\alpha$ and $X \subseteq \alpha$, let $\pi_r : (field(r), r) \rightarrow (\alpha, \leq)$ be the canonical isomorphism and set $Code(X, r) = \pi_r^{-1}(X)$. Note that if $M$ is a transitive set and $r \in M$, then $\pi_r \in M$. Thus, for transitive set $M$ with $r \in M$, we have

$$X \in M \Leftrightarrow Code(X, r) \in M.$$

Now we define a $\Sigma^1_1$ set of reals and show that it has property $\bigstar$. For $x, y \in \omega^\omega$, $x \leq_h y$ stands for: $x$ is hyperarithmetic in $y$. The reals in $L_{\omega_1^x}[x]$ are precisely the reals hyperarithmetic in $x$.

**Definition A.13** (Sami, [1] ) For $a, b \in \omega^\omega$, define $a \sqsubseteq b \Leftrightarrow \forall x \leq_h a (x \leq_T b) \wedge \omega_1^a = \omega_1^b$. Define $D = \{x \subseteq \omega \mid \exists y (y \sqsubseteq x)\}$.

**Fact A.14** ([2]) $\{(x, y) \in \omega^\omega \times \omega^\omega \mid \omega_1^x \leq \omega_1^y\}$ *is* $\Sigma^1_1$ *but not* $\Pi^1_1$.

By Fact A.14, $D$ is $\Sigma^1_1$. Now we show that $D$ has property $\bigstar$.

**Theorem A.15** $D$ *has property* $\bigstar$.

*Proof* Note that $D$ is Turing-closed. In the following, we show that $D$ is cofinal in the Turing degrees and satisfies condition (2) in Definition A.1.

**Lemma A.2** $D$ *is cofinal in the Turing degrees.*

*Proof* We show that $\forall a \in \omega^\omega \exists b \in D(a \leq_T b)$. Let

$$A = \{y \in \omega^\omega \mid \forall x \leq_h a(x \leq_T y)\}.$$

Note that $A$ is $\Sigma^1_1(a)$ and $A \neq \emptyset$ (Since $\{x \subseteq \omega \mid x \leq_h a\}$ is countable, let $\{x_n : n \in \omega\}$ be a list of such $x$ and $y = \oplus_{n \in \omega} x_n$, then $y \in A$). By Gandy basis theorem (cf. Fact A.10), there is $b \in A$ such that $a \leq_T b$ and $\omega_1^a = \omega_1^b$. Thus, $a \sqsubseteq b$ and $b \in D$. $\qquad \square$

**Lemma A.3** (Sami, [1]) *Suppose* $a \in D, \alpha < \omega_1^a$ *and* $r \in \mathsf{WO}_\alpha$. *If* $X \in \mathscr{P}(\alpha) \cap L_{\omega_1^a}$, *then* $Code(X, r)$ *is* $\Sigma_{\alpha+2}^0(a, r)$.

*Proof* Let $b \sqsubseteq a$. Since $\omega_1^b = \omega_1^a$ and $\alpha < \omega_1^a$, we have $\alpha < \omega_1^b$. By Fact A.12, pick $s \in \mathsf{WO}_\alpha$ such that $s \leq_T b$ and $(\omega, s) \cong (\omega, r)$ (since $\alpha < \omega_1^b$, there is $r' \in \mathsf{WO}_\alpha$ such that $r' \leq_T b$. For such $r'$, there is $s \leq_T r'$ such that $(\omega, r) \cong (\omega, s)$. So such $s$ exists). Let $x = Code(X, r)$ and $y = Code(X, s)$. Since $s \leq_T b$, we have $s \in L_{\omega_1^b}[b]$. Since $X, s \in L_{\omega_1^b}[b]$, we have $y \in L_{\omega_1^b}[b]$. Thus, $y \leq_h b$. Since $b \sqsubseteq a$, we have $y \leq_T a$. For $k \in \omega$, note that

$$k \in x \Leftrightarrow \exists k'(k' \in y \wedge [(k \in field(r) \leftrightarrow k' \in field(s)) \wedge s \restriction k' \in \mathsf{WO}_{|r \restriction k|}]).$$

Since $y \leq_T a$, we have "$k' \in y$" is a $\Sigma_1^0(a)$ property of $k'$. Since $s \leq_T b$ and $b \sqsubseteq a$, we have $s \leq_T a$. Thus, by Proposition A.2, $x$ is $\Sigma_{\alpha+2}^0(a, r)$. $\square$

**Lemma A.4** (Sami, [1]) *Suppose* $C$ *is a Turing cone with* $C \subseteq D$, $c \in C$, $\xi < \omega_1^c$, $X \subseteq \xi$ *and* $X \in L$. *Then* $X \in L_{\omega_1^c}[c]$.

*Proof* Since $\xi < \omega_1^c$, take $r \in \mathsf{WO}_\xi$ such that $r \leq_T c$. Suppose $X \subseteq \xi$ and $X \in L$. Let $X \in L_\delta$ for some $\delta < \omega_1$. Pick any $s \in \mathsf{WO}_\delta$. Since $\delta < \omega_1^s \leq \omega_1^{c \oplus s}$, we have $X \in L_{\omega_1^{c \oplus s}}$. Also $c \oplus s \in Cone(c) \subseteq D$. Since $\xi < \omega_1^c \leq \omega_1^{c \oplus s}$, $X \subseteq \xi$ and $X \in L_{\omega_1^{c \oplus s}}$, by Lemma A.3, we have $Code(X, r)$ is $\Sigma_{\xi+2}^0(c \oplus s, r)$.[2] Since $r \leq_T c$, we have $Code(X, r)$ is $\Sigma_{\xi+2}^0(c \oplus s)$. Theorem A.11 relativized to $c$ gives that:

if $a \in \omega^\omega$, $1 \leq \alpha < \omega_1^c$ and for some $\eta < \omega_1$, $a$ is $\Sigma_\alpha^0(x \oplus c)$ for all $x \in \mathsf{WO}_\eta$, then $a$ is $\Sigma_\alpha^0(c)$.

Since $Code(X, r)$ is $\Sigma_{\xi+2}^0(c \oplus s)$ for any $s \in \mathsf{WO}_\delta$, by Theorem A.11, we have $Code(X, r)$ is $\Sigma_{\xi+2}^0(c)$. Hence $Code(X, r) \in L_{\omega_1^c}[c]$. Since $r \leq_T c$, we have $X \in L_{\omega_1^c}[c]$. $\square$

By Lemmas A.2 and A.4, $D$ has property $\bigstar$, and we are done. $\square$

We assume $Z_2 + Det(Turing-\Sigma_1^1)$. Since $D$ is Turing closed and cofinal in the Turing degrees, $D$ contains a cone of Turing degrees $Cone(c)$ for some $c \in \omega^\omega$. Since $D$ has property $\bigstar$, by Theorem A.3, if $\alpha$ is a countable $c$-admissible ordinal, then $\alpha$ is an $L$-cardinal (i.e. **HP** holds).

---

[2]In Lemma A.3, let $a = c \oplus s$ and $\alpha = \xi$.

# Appendix B
# $Det\,(<\omega^2\text{-}\Pi_1^1)$ Implies that $0^\sharp$ Exists in $\mathbf{Z}_3$

**Abstract** In this appendix, I reconstruct Martin's proof of "$Det\,(<\omega^2\text{-}\Pi_1^1)$ implies that $0^\sharp$ exists" in $\mathbf{Z}_3$ *without* the use of Harrington's Principle.

First of all, besides $Det\,(\Sigma_1^1)$, there are other determinacy hypotheses equivalent to the existence of $0^\sharp$ in ZF. Examples include: $Det(\text{Turing-}\Sigma_1^1)$, $Det\,(\Sigma_1^1\text{-Wadge})$, and $Det\,(\Sigma_1^1\text{-Kleene})$. For the definition of $Det\,(<\omega^2\text{-}\Pi_1^1)$, see Definition B.1. For the definition of $Det\,(\Sigma_1^1\text{-Wadge})$ and $Det\,(\Sigma_1^1\text{-Kleene})$, I refer to [8, 9].

Now, Harrington first proved in [8] that $Det\,(\Sigma_1^1\text{-Wadge})$ is equivalent to "$0^\sharp$ exists". G. Weitkamp proved in [9] that $Det\,(\Sigma_1^1\text{-Kleene})$ is equivalent to "$0^\sharp$ exists". Martin proved that $Det\,(<\omega^2\text{-}\Pi_1^1)$ is equivalent to "$0^\sharp$ exists" (see [10]).

In this chapter, I list some known facts about these determinacy hypotheses. Let $\Gamma$ be one of the following statements: $Det(\text{Turing-}\Sigma_1^1)$, $Det\,(\Sigma_1^1\text{-Wadge})$ and $Det\,(\Sigma_1^1\text{-Kleene})$. It is provable in $\mathbf{Z}_2$ that $0^\sharp$ exists implies $\Gamma$. All known proofs of "$\Gamma$ implies that $0^\sharp$ exists" are done in two steps: first show that $\Gamma$ implies HP and then show that HP implies that $0^\sharp$ exists. The first step is provable in $\mathbf{Z}_2$. The minimal system in higher-order arithmetic to show that HP implies that $0^\sharp$ exists is $\mathbf{Z}_4$. As a corollary, it is provable in $\mathbf{Z}_4$ that $\Gamma$ is equivalent to $0^\sharp$ exists.

However, among the determinacy hypotheses which are equivalent to "$0^\sharp$ exists", one exception is $Det\,(<\omega^2\text{-}\Pi_1^1)$. Martin's proof of "$Det\,(<\omega^2\text{-}\Pi_1^1)$ implies $0^\sharp$ exists" does not use Harrington's Principle. I have previously observed that Martin's proof can be done in $\mathbf{Z}_3$. In this Appendix, I reconstruct Martin's proof of "$Det\,(<\omega^2\text{-}\Pi_1^1)$ implies that $0^\sharp$ exists" in $\mathbf{Z}_3$ without the use of Harrington's Principle.

**Definition B.1** ([6])

(1) Suppose $\alpha$ is recursive and there is a recursive well-ordering $E \subseteq \omega^\omega \times \omega^\omega$ such that $o.t.(E) = \alpha$. For $n \in \omega$, let $|n|$ denote the order type of the predecessors of $n$ according to $E$. For $A \subseteq \omega^\omega$, $A$ is $\alpha\text{-}\Pi_1^1$ if there exists a sequence $\langle A_\xi \mid \xi < \alpha\rangle$ of subsets of $\omega^\omega$ such that $\{(n, x) \in \omega \times \omega^\omega \mid x \in A_{|n|}\} \in \Pi_1^1$ and $x \in A$ if the

Y. Cheng, *Incompleteness for Higher-Order Arithmetic*, SpringerBriefs in Mathematics, https://doi.org/10.1007/978-981-13-9949-7

least $\xi$ such that $\xi = \alpha$ or $x \notin A_\xi$ is odd.[3] In this case, we say that $\langle A_\xi \mid \xi < \alpha \rangle$ witnesses $A \in \alpha\text{-}\Pi_1^1$.

(2) $A$ is $< \alpha\text{-}\Pi_1^1$ if it is $\beta\text{-}\Pi_1^1$ for some $\beta < \alpha$.

**Theorem B.2** (Martin, [10], $Z_3$) $Det(<\omega^2\text{-}\Pi_1^1)$ *implies that* $0^\sharp$ *exists.*

*Proof* From Proposition 2.5, $0^\sharp$ exists iff $L_{\omega_1}$ has an uncountable set of indiscernibles. Assume $Det(<\omega^2\text{-}\Pi_1^1)$. It suffices to show that $L_{\omega_1}$ has an uncountable set of indiscernibles as follows.

**Lemma B.1** *For any formula* $\varphi(x_0, \ldots, x_{n-1})$ *in* $\mathfrak{L}_{st}$, *there exists a closed and unbounded subset* $C_\varphi$ *of* $\omega_1$ *such that for any two increasing sequences* $\langle \rho_i \mid i \leq n \rangle$ *and* $\langle \theta_i \mid i \leq n \rangle$ *from* $C_\varphi$ *we have*

$$L_{\rho_n} \models \varphi[\rho_0, \ldots, \rho_{n-1}] \Leftrightarrow L_{\theta_n} \models \varphi[\theta_0, \ldots, \theta_{n-1}].$$

*Proof* Fix a formula $\varphi(x_0, \ldots, x_{n-1})$. We will design a game $G$ which is $< \omega^2\text{-}\Pi_1^1$ and hence determined. Let $\pi : \omega \to \omega \cdot (n+1)$ be a recursive function such that for any $\beta < \omega \cdot (n+1)$, $\pi^{-1}(\beta)$ is infinite. For each $\beta < \omega \cdot (n+1)$, let $\pi_\beta : \omega \to \pi^{-1}(\beta)$ be bijection defined by $\pi_\beta(j)$ be the least element of $\pi^{-1}(\beta) \setminus \{\pi_\beta(i) : i < j\}$. If $x$ is a play of the game and $\beta < \omega \cdot (n+1)$, let $((x)_I)_\beta$ be the real $x \upharpoonright \{2 \cdot \pi_\beta(k) : k \in \omega\}$; $((x)_{II})_\beta$ be the real $x \upharpoonright \{2 \cdot \pi_\beta(k) + 1 : k \in \omega\}$.

Let $S_\beta(x)$ and $T_\beta(x)$ be the binary relation on $\omega$ coded respectively by the real $((x)_I)_\beta$ and $((x)_{II})_\beta$. Now we define the game $G$. Given $x$ which is a play of the game, player I loses the game if for some $\beta < \omega \cdot (n+1)$, $S_\beta(x)$ is not a well-ordering but $T_\eta(x)$ is a well-ordering for any $\eta < \beta$; player II loses the game if for some $\beta < \omega \cdot (n+1)$, $T_\beta(x)$ is not a well-ordering but $S_\eta(x)$ is a well-ordering for any $\eta < \beta$. Let $\gamma_\beta(x)$ be the order type of $S_\beta(x)$ if $S_\beta(x)$ is a well-ordering on $\omega$; let $\delta_\beta(x)$ be the order type of $T_\beta(x)$ if $T_\beta(x)$ is a well-ordering on $\omega$. If $\gamma_\beta(x)$ and $\delta_\beta(x)$ are defined for all $\beta < \omega \cdot (n+1)$, let

$$\theta_i = \theta_i(x) = \sup(\{\gamma_{\omega \cdot i + k}(x), \delta_{\omega \cdot i + k}(x) \mid k \in \omega\})$$

for all $i \leq n$. If $\gamma_\beta(x)$ and $\delta_\beta(x)$ are defined for all $\beta < \omega \cdot (n+1)$, then player one wins $G$ iff

$$L_{\theta_n} \models \varphi[\theta_0, \ldots, \theta_{n-1}].$$

Thus $G$ is a $\omega \cdot (n+2)\text{-}\Pi_1^1$ game and hence has a winning strategy $\tau$.

**Proposition B.1** *There exists a closed and unbounded set* $C$ *of* $\omega_1$ *such that for any increasing sequence* $\langle \rho_i \mid i \leq n \rangle$ *of elements from* $C$, *there exists a play* $x$ *which is consistent with* $\tau$ *such that for any* $i \leq n$, $\theta_i(x)$ *are defined and* $\theta_i(x) = \rho_i$.

*Proof* We assume that $\tau$ is a winning strategy for player I. The same argument applies to the case $\tau$ is a winning strategy for player II.

---

[3] An odd ordinal is of the form $\gamma + 2n + 1$ for some limit ordinal $\gamma$ and $n \in \omega$.

If $\eta < \omega_1$ and $\beta < \omega \cdot (n+1)$, let $C_\eta^\beta = \{y : \exists z \in \omega^\omega$ such that $y = \tau * z$ and $\forall \beta' < \beta(\delta_{\beta'}(y)$ is defined and $\delta_{\beta'}(y) < \eta)\}$. Note that $C_\eta^\beta$ is $\Sigma_1^1$. Since player I wins the game, for any $y \in C_\eta^\beta$, we have $S_\beta(y) \in$ WO and $\gamma_\beta(y)$ is defined. By $\Sigma_1^1$ boundedness theorem (cf. Fact A.6), we have $\sup(\{\gamma_\beta(y) \mid y \in C_\eta^\beta\}) < \omega_1$. Let $v(\beta, \eta) = \sup(\{\gamma_\beta(y) \mid y \in C_\eta^\beta\})$ and $v(\eta) = \sup(\{v(\beta, \eta) \mid \beta < \omega \cdot (n+1)\})$. Note that $v(\eta) < \omega_1$ for any $\eta < \omega_1$. Let

$$C = \{\alpha < \omega_1 \mid \alpha \text{ is limit ordinal and for any } \eta < \alpha, v(\eta) < \alpha\}.$$

It is easy to check that $C$ is club subset of $\omega_1$. Let $\langle \rho_i \mid i \le n \rangle$ be a strictly increasing sequence from $C$. For any $i \le n$, let $\langle \xi_{\omega \cdot i + m} : m \in \omega \rangle$ be an increasing sequence converging to $\rho_i$. Given $i \le n, m \in \omega$, let $z_{i,m}$ be the real which codes a well-ordering on $\omega$ with order type $\xi_{\omega \cdot i + m}$. Let player II play $y \subseteq \omega$ such that $((\tau * y)_{II})_{\omega \cdot i + m} = z_{i,m}$. Let $x = \tau * y$. We will show that $x$ is the play we want.

Note that for any $i \le n, m \in \omega$, $T_{\omega \cdot i + m}(x)$ is a well-ordering on $\omega$ coded by $z_{i,m}$, $\delta_{\omega \cdot i + m}(x)$ is defined and $\delta_{\omega \cdot i + m}(x) = \xi_{\omega \cdot i + m}$. Thus, $\delta_\beta(x)$ is defined for all $\beta < \omega \cdot (n+1)$. Since player I wins the game, $\gamma_\beta(x)$ is defined for all $\beta < \omega \cdot (n+1)$. Thus, for all $\beta < \omega \cdot (n+1)$, $\delta_\beta(x)$ and $\gamma_\beta(x)$ are defined. Hence for any $i \le n$, $\theta_i(x)$ is defined.

Now we show that for any $i \le n$, we have $\theta_i(x) = \rho_i$. Fix $i \le n$. Since for all $m \in \omega$, $\delta_{\omega \cdot i + m}(x) = \xi_{\omega \cdot i + m}$, by definition, we have $\theta_i(x) \ge \sup(\{\xi_{\omega \cdot i + m} : m \in \omega\}) = \rho_i$. To show that $\theta_i(x) = \rho_i$, it suffices to show that for any $m \in \omega$, we have $\gamma_{\omega \cdot i + m}(x) < \rho_i$. Given $m \in \omega$, since $\rho_i \in C$ and $\rho_i$ is a limit ordinal, there is $\zeta < \rho_i$ such that for all $\beta' < \omega \cdot i + m$, $\delta_{\beta'}(x)$ is defined and $\delta_{\beta'}(x) < \zeta$. Such $\zeta$ exists since $\rho_{i-1} < \rho_i$ and both are limit ordinals. Thus, by definition, we have $x \in C_\zeta^{\omega \cdot i + m}$. Hence $\gamma_{\omega \cdot i + m}(x) \le v(\zeta)$. By definition of $C$, since $\rho_i \in C$ and $\zeta < \rho_i$, we have $v(\zeta) < \rho_i$. Thus we have shown that $\gamma_{\omega \cdot i + m}(x) < \rho_i$ for any $m \in \omega$. Hence $\theta_i(x) = \rho_i$.

Take $C$ be the club subset of $\omega_1$ as in Proposition B.1. We show that for any increasing sequence $\langle \rho_i \mid i \le n \rangle$ from $C$, if $\tau$ is a winning strategy for player I, then

$$L_{\rho_n} \models \varphi[\rho_0, \ldots, \rho_{n-1}];$$

if $\tau$ is a winning strategy for player II, then

$$L_{\rho_n} \not\models \varphi[\rho_0, \ldots, \rho_{n-1}].$$

Let $\langle \rho_i \mid i \le n \rangle$ be any increasing sequence from $C$. By Proposition B.1, there exists a play $x$, consistent with $\tau$, such that for any $i \le n$, $\theta_i(x)$ are defined and $\theta_i(x) = \rho_i$. If $\tau$ is a winning strategy for player I, then $x$ is a win for player I and, since all $\theta_i(x)$ are defined, $L_{\rho_n} \models \varphi[\rho_0, \ldots, \rho_{n-1}]$. Similarly, if $\tau$ is a winning strategy for player II, then since all $\theta_i(x)$ are defined, $L_{\rho_n} \not\models \varphi[\rho_0, \ldots, \rho_{n-1}]$.

Let $C_\varphi = C$. Given any two increasing sequences $\langle \rho_i \mid i \le n \rangle, \langle \theta_i \mid i \le n \rangle$ from $C_\varphi$, if $\tau$ is a winning strategy in $G$ for player I, we have $L_{\rho_n} \models \varphi[\rho_0, \ldots, \rho_{n-1}]$ and $L_{\theta_n} \models \varphi[\theta_0, \ldots, \theta_{n-1}]$; if $\tau$ is a winning strategy in $G$ for player II, we have $L_{\rho_n} \not\models$

$\varphi[\rho_0, \ldots, \rho_{n-1}]$ and $L_{\theta_n} \nvDash \varphi[\theta_0, \ldots, \theta_{n-1}]$. Thus, for any two increasing sequences $\langle \rho_i \mid i \le n \rangle$ and $\langle \theta_i \mid i \le n \rangle$ from $C_\varphi$, we have

$$L_{\rho_n} \vDash \varphi[\rho_0, \ldots, \rho_{n-1}] \Leftrightarrow L_{\theta_n} \vDash \varphi[\theta_0, \ldots, \theta_{n-1}]$$

and we are done.                                                                                          □

For any formula $\varphi$ in $\mathfrak{L}_{st}$, let $C_\varphi$ be the club subset of $\omega_1$ as in Lemma B.1. Let $C_0$ be the intersection of $C_\varphi$ for any formula $\varphi$ in $\mathfrak{L}_{st}$. Since there are only countable many formulas, $C_0$ is a club subset of $\omega_1$. Let

$$C_1 = \{\alpha < \omega_1 : L_\alpha \prec L_{\omega_1}\}.$$

Then $C_1$ is a club subset of $\omega_1$. Let $C = C_0 \cap C_1$. $C$ is a club subset of $\omega_1$.

**Claim** *$C$ is a set of indiscernibles for $L_{\omega_1}$.*

*Proof* For any formula $\varphi(x_0, \ldots, x_{n-1})$ in $\mathfrak{L}_{st}$, we show that for any two increasing sequences $\langle \rho_i \mid i < n \rangle$ and $\langle \theta_i \mid i < n \rangle$ from $C$ we have

$$L_{\omega_1} \vDash \varphi[\rho_0, \ldots, \rho_{n-1}] \Leftrightarrow L_{\omega_1} \vDash \varphi[\theta_0, \ldots, \theta_{n-1}].$$

Take $\rho_n$ and $\theta_n$ from $C$ such that $\rho_n > \rho_{n-1}$ and $\theta_n > \theta_{n-1}$. Since $\langle \rho_i \mid i \le n \rangle$ and $\langle \theta_i \mid i \le n \rangle$ are from $C_\varphi$, by Lemma B.1, we have

$$L_{\rho_n} \vDash \varphi[\rho_0, \ldots, \rho_{n-1}] \Leftrightarrow L_{\theta_n} \vDash \varphi[\theta_0, \ldots, \theta_{n-1}].$$

Since $\rho_n$ and $\theta_n$ are from $C_1$, we have $L_{\rho_n} \prec L_{\omega_1}$ and $L_{\theta_n} \prec L_{\omega_1}$. So

$$L_{\rho_n} \vDash \varphi[\rho_0, \ldots, \rho_{n-1}] \Leftrightarrow L_{\omega_1} \vDash \varphi[\rho_0, \ldots, \rho_{n-1}]$$

and

$$L_{\theta_n} \vDash \varphi[\theta_0, \ldots, \theta_{n-1}] \Leftrightarrow L_{\omega_1} \vDash \varphi[\theta_0, \ldots, \theta_{n-1}].$$

Hence we have

$$L_{\omega_1} \vDash \varphi[\rho_0, \ldots, \rho_{n-1}] \Leftrightarrow L_{\omega_1} \vDash \varphi[\theta_0, \ldots, \theta_{n-1}].$$

This finishes the proof of the claim.                                                                    □

This finishes the proof of Theorem B.2.                                                                   □

*Remark B.1* Martin's proof of "$Det(<\omega^2\text{-}\Pi_1^1)$ implies that $0^\sharp$ exists" does not use HP. But all known proofs of "$Det(\Pi_1^1)$ implies that $0^\sharp$ exists" use HP.

Martin first proved in $\mathsf{ZF}$ that $0^\sharp$ exists implies $Det(<\omega^2\text{-}\Pi_1^1)$ (cf. [10]). I have checked that it is in fact provable in $\mathsf{Z}_2$. The following theorem is an observation from Martin's original proof of "$0^\sharp$ exists implies $Det(< \omega^2\text{-}\Pi_1^1)$" in ZF. For details about Martin's proof, I refer to [10].

**Theorem B.3** (Martin, [10], $Z_2$) *If $0^\sharp$ exists and $A \subseteq \omega^\omega$ is in $< \omega^2\text{-}\Pi_1^1$, then $G_A$ has a winning strategy in $L[0^\sharp]$. Thus, $Z_2 + 0^\sharp$ exists implies $Det(< \omega^2\text{-}\Pi_1^1)$.*

From Theorems B.3 and B.2, $Det(< \omega^2\text{-}\Pi_1^1)$ is equivalent to $0^\sharp$ exists in $Z_3$. As far as I know, the question "whether $Z_2 + Det(< \omega^2\text{-}\Pi_1^1)$ implies that $0^\sharp$ exists" is open.

# Appendix C
# Other Notions of Large Cardinals

**Abstract** In this appendix, I review some notions of large cardinals used in this book which are not covered in Sects. 2.1.2 and 2.1.3.

The large cardinal notions used in this book are mainly large cardinals compatible with $L$. The following notions of large cardinals are compatible with $L$: inaccessible cardinal, reflecting cardinal, Mahlo cardinal, weakly compact, indescribable cardinal, unfoldable cardinal, subtle cardinal, ineffable cardinal, remarkable cardinal, $\alpha$-iterable cardinal$\alpha$-iterable cardinal for $\alpha < \omega_1^L$ and $\alpha$-Erdös cardinal for $\alpha < \omega_1$.

The hierarchy of large cardinals compatible with $L$ in terms of consistency strength is as follows: $\alpha$-Erdös cardinal $(\omega < \alpha < \omega_1)$ > $\omega$-Erdös cardinal > $n$-iterable cardinal $(2 < n \in \omega)$ > 2-iterable cardinal > remarkable cardinal > 1-iterable cardinal > totally ineffable cardinal > $n$-ineffable cardinal > $n$-subtle cardinal > unfoldable cardinal > totally indescribable cardinal > $\Pi_m^n$-indescribable > weakly compact > Mahlo cardinal > reflecting cardinal > inaccessible cardinal.

For the definition of remarkable cardinals, I refer to Sect. 2.1.3. In the following, I give definitions for the other large cardinal notions used in this book.

**Definition C.1** ([5]) For an uncountable cardinal $\kappa$, $\kappa$ is weakly inaccessible if it is a regular limit cardinal; $\kappa$ is strong limit if $2^\lambda < \kappa$ for any $\lambda < \kappa$; $\kappa$ is inaccessible if it is a regular strong limit cardinal.

Under **GCH**, a cardinal is inaccessible iff it is weakly inaccessible. Under **AC**, if $\kappa$ is regular, then $\forall x(x \in H_\kappa \leftrightarrow (x \subseteq H_\kappa \wedge |x| < \kappa))$. Under **AC**, if $\kappa > \omega$ is regular, then $H_\kappa \models \mathbf{ZFC}^-$.

**Proposition C.1** ([11], **AC**) *Let $\kappa$ be an uncountable regular cardinal. Then the following are equivalent: (1) $H_\kappa \models \mathbf{ZFC}$; (2) $H_\kappa = V_\kappa$; (3) $\kappa$ is inaccessible.*

Now I introduce the notion of reflecting cardinal and Mahlo cardinal.

**Definition C.2** ([12]) A regular cardinal $\kappa$ is reflecting if for every $a \in H_\kappa$ and every first-order formula $\varphi(x)$, if for some regular cardinal $\lambda$, $H_\lambda \models \varphi(a)$, then there exists a cardinal $\delta < \kappa$ such that $H_\delta \models \varphi(a)$.

© The Author(s), under exclusive license to Springer Nature Singapore Pte Ltd. 2019

Y. Cheng, *Incompleteness for Higher-Order Arithmetic*, SpringerBriefs in Mathematics, https://doi.org/10.1007/978-981-13-9949-7

Note that if $\kappa$ is reflecting, then $\kappa$ is inaccessible.

**Definition C.3** ([6]) An uncountable cardinal $\kappa$ is Mahlo if $\kappa$ is inaccessible and $\{\alpha < \kappa \mid \alpha \text{ is inaccessible}\}$ is stationary in $\kappa$.

If $\kappa$ is Mahlo, then $\kappa$ is the $\kappa$-th inaccessible cardinal.

Next, I introduce the notion of weakly compact cardinal (cf. [5]). We consider infinitary languages $\mathscr{L}_{\kappa,\lambda}$ for infinite cardinal $\kappa$, $\lambda$, which are generalizations of the ordinary first-order language. An $\mathscr{L}_{\kappa,\lambda}$ language is formulated as follows: as the usual first-order logic, first specify a supply of non-logical symbols: finitary relation, function and constant symbols. These together with an allowed supply of $\max\{\kappa, \lambda\}$ many variables lead to the terms and atomic formulas. Then the usual formula generating rules are expanded to allow conjunction $\bigwedge_{\xi<\alpha}$ and disjunction $\bigvee_{\xi<\alpha}$ of $\alpha$ many formulas for any $\alpha < \kappa$, and universal quantification $\forall_{\xi<\beta}$ and existential quantification $\exists_{\xi<\beta}$ of $\beta$ many variables for any $\beta < \lambda$. Finally, a formula is an expression so generated with less than $\lambda$ free variables, this to allow the possibility of quantification closure. Structures for interpreting the language are as for first-order logic, and the satisfaction relation is extended to incorporate the new infinitary connectives and quantifiers in the expected way.

A set of $\mathscr{L}_{\kappa,\lambda}$ sentences $\Sigma$ is satisfiable if $\Sigma$ has a model under the expected interpretation of infinitary conjunction, disjunction and quantification; $\Sigma$ is $\theta$ satisfiable if for every $S \subseteq \Sigma$ with $|S| < \theta$ is satisfiable.[4]

A tree is a partially ordered set $(T, \leq_T)$ such that for all $t \in T$, $\{s \in T \mid s \leq_T t\}$ is well-ordered by $\leq_T$ and $T$ has a unique minimal element, the root of the tree. A branch through a tree $(T, \leq_T)$ is a linearly ordered subset and its length is the order type of the linear order.

**Definition C.4** ([5, 6]) Let $\kappa > \omega$ and $\lambda \geq \omega$.

(1) $\kappa$ is *weakly compact* if for any set of $\mathscr{L}_{\kappa,\kappa}$ sentences $\Sigma$ with $|\Sigma| \leq \kappa$, if $\Sigma$ is $\kappa$-satisfiable, then $\Sigma$ is satisfiable.[5]
(2) A regular cardinal $\kappa$ has the *tree property* if every $\kappa$-tree has a $\kappa$ branch.
(3) $\kappa$ has the *Extension property* if for any $R \subseteq V_\kappa$ there is a transitive set $X \supseteq V_\kappa$ with $\kappa \in X$ and an $S \subseteq X$ such that $\langle V_\kappa, \in, R \rangle \prec \langle X, \in, S \rangle$.
(4) $\kappa$ has the *linear order property* if whenever $(L, <_L)$ is a linear ordering of cardinality $\geq \kappa$, there is a monotone $<_L$ sequence of order type $\kappa$.

I now introduce the notion of indescribable cardinal. We first define $\Pi_m^n$ formula ($\Sigma_m^n$ formula) (cf. [5]). Let $n > 0$ be a natural number and let us consider the $n$-th order predicate calculus. There are variables of orders $1, 2, \ldots, n$, and the quantifiers are applied to variables of all orders. An $n$-th order formula contains, in addition to

---

[4]The language $\mathscr{L}_{\omega,\omega}$ is just the language of the first-order predicate calculus. $\mathscr{L}_{\omega,\omega}$ satisfies the Compactness Theorem: if $\Sigma$ is a set of sentences such that every finite $S \subseteq \Sigma$ has a model, then $\Sigma$ has a model.

[5]$|\Sigma| \leq \kappa$ is equivalent to $\Sigma$ has at most $\kappa$ many non-logical symbols.

first-order symbols and higher-order quantifiers, predicates $X(z)$ where $X$ and $z$ are variables of order $k + 1$ and $k$ respectively (for any $k < n$).

Satisfaction for an $n$-th order formula in a model $\mathfrak{A} = (A, P, \ldots, f, \ldots, c \cdots)$ is defined as follows: variables of first-order are interpreted as elements of the set $A$, variables of second-order as elements of $\mathscr{P}(A)$ etc.; variables of order n are interpreted as elements of $\mathscr{P}^{n-1}(A)$. The predicate $X(z)$ is interpreted as $z \in X$. A $\Pi_m^n$ formula is a formula of order $n + 1$ of the form

$$\underbrace{(\forall X)(\exists Y) \cdots \psi}_{m \text{quantifiers}} \tag{C.1}$$

where $X, Y, \ldots$ are $(n + 1)$-th order variables and $\psi$ is such that all quantified variables are of order at most n. Similarly, a $\Sigma_m^n$ formula is as in (C.1), but with $\exists$ and $\forall$ interchanged.

### Definition C.5 ([5, 6])

(1) For $Q$ either $\Pi_n^m$ or $\Sigma_n^m$, $\kappa$ is $Q$-indescribable if for any $R \subseteq V_\kappa$ and $Q$ sentence $\varphi$ such that $\langle V_\kappa, \in, R \rangle \models \varphi$, there is an $\alpha < \kappa$ such that $\langle V_\alpha, \in, R \cap V_\alpha \rangle \models \varphi$.
(2) $\kappa$ is *totally indescribable* if $\kappa$ is $\Pi_n^m$-indescribable for any $m, n \in \omega$.

### Definition C.6
If $f$ is a function on $[X]^n$, then we say that $Y \subseteq X$ is *homogeneous* for $f$ if $f$ is constant on $[Y]^n$. $\kappa \to (\lambda)_\xi^n$ means that for any $f : [\kappa]^n \to \xi$, there is a $X \subseteq \kappa$ such that $|X| = \lambda$ and $X$ is homogeneous for $f$. $\kappa \to (\alpha)_\lambda^{<\omega}$ means that for any $f : [\kappa]^{<\omega} \to \lambda$, there is $X \subseteq \kappa$ such that $X$ has order type $\alpha$ and for each $n \in \omega$, $X$ is homogeneous for $f \upharpoonright [\kappa]^n$.

As a summary, we have the following equivalent characterizations of weakly compact cardinals.

### Theorem C.7 ([5, 6, 19]) *For uncountable cardinal $\kappa$, the following are equivalent:*

*(1) $\kappa$ is weakly compact.*
*(2) $\kappa$ is $\Pi_1^1$-indescribable.*
*(3) $\kappa \to (\kappa)_\lambda^n$ for every $n \in \omega, \lambda < \kappa$.*
*(4) $\kappa$ is inaccessible and has the tree property.*
*(5) $\kappa$ has the Extension property.*
*(6) $\kappa$ has the linear order property.*

Finally, I introduce the notion of unfoldable cardinal. For definitions and facts about unfoldable cardinal, I refer to [13–15].

### Definition C.8
Let $(M, E)$ and $(N, F)$ be models of set theory. We say $(N, F)$ *end extends* $(M, E)$ if for every $a \in M$, $\{b \in M \mid bEa\} = \{b \in N \mid bFa\}$. Then $(M, E) \prec_e (N, F)$ means that $(N, F)$ is an elementary end extension of $(M, E)$. The structure $(\mathfrak{F}_M, \prec_e)$ consists of all non-trivial elementary end extension of $M$, ordered by the relation $\prec_e$.

**Definition C.9** ([13–15]) An inaccessible cardinal $\kappa$ is unfoldable iff for any $S \subseteq \kappa$ and any $\alpha$, there exist transitive set $N$ and $S' \subseteq N$ such that $(V_\kappa, \in, S) \prec_e (N, \in, S')$ (i.e. $N \in (\mathfrak{F}_{(V_\kappa, \in, S)}, \prec_e)$) and $o(N) \geq \alpha$, where $o(N) = \mathsf{Ord} \cap N$.

The notions subtle and ineffable cardinals are introduced next. For definitions and facts about subtle ineffable cardinals, I refer to [16, 17].

**Definition C.10** ([16, 17])

(1) Let $A$ be a set of ordinals and $1 \leq n < \omega$. A sequence $\mathbf{S} = \langle S_{\beta_1,\ldots,\beta_n} \mid \beta_1 < \cdots < \beta_n, \beta_1, \ldots, \beta_n \in A\rangle$ is an $(n, A)$ sequence iff $S_{\beta_1,\ldots,\beta_n} \subseteq \beta_1$ for all $\beta_1, \ldots, \beta_n$ in $A$. In particular, it makes sense to speak of an $(n, \kappa)$ sequence.
(2) We say that $X$ is *homogeneous for an $(n, A)$ sequence* $\mathbf{S}$ if $X \subseteq A$ and for any two sequences $\beta_1 < \cdots < \beta_n, \beta'_1 < \cdots < \beta'_n$ from $X$, if $\beta_1 \leq \beta'_1$, then $S_{\beta_1,\ldots,\beta_n} = \beta_1 \cap S_{\beta'_1,\ldots,\beta'_n}$.
(3) A cardinal $\kappa$ is *$n$-subtle* if for every $(n, \kappa)$ sequence $\mathbf{S}$ and every closed unbounded set $C \subseteq \kappa$, there is $x \in [C]^{n+1}$ such that $x$ is homogeneous for $\mathbf{S}$.
(4) $\kappa$ is a *subtle cardinal* if $\kappa$ is 1-subtle. i.e. $\kappa$ is subtle if for any sequence $\langle S_\alpha \mid \alpha < \kappa\rangle$ with $S_\alpha \subseteq \alpha$ and any club $C$ on $\kappa$, there exists $\alpha, \beta$ in $C$ such that if $\alpha < \beta$, then $S_\alpha = S_\beta \cap \alpha$.

**Definition C.11** ([16, 17])

(1) $\kappa$ is *$n$-ineffable* if every $(n, \kappa)$ sequence has a homogenous set which is stationary in $\kappa$.
(2) $\kappa$ is *ineffable* if $\kappa$ is 1-ineffable. i.e. for any sequence $\langle S_\alpha \mid \alpha < \kappa\rangle$ with $S_\alpha \subseteq \alpha$, there exists $A \subseteq \kappa$ such that $A$ is stationary and for all $\alpha, \beta \in A$, if $\alpha < \beta$, then $S_\alpha = S_\beta \cap \alpha$.

The notion of $\alpha$-iterable cardinal$\alpha$-iterable cardinal is introduced next. For definitions and facts about $\alpha$-iterable cardinal$\alpha$-iterable cardinal, I refer to [18].

To define $\alpha$-iterable cardinals$\alpha$-iterable cardinals, we will need the key notion of $\alpha$-good $M$-ultrafilters.

**Definition C.12**

(1) A weak $\kappa$-model $M$ of set theory is a transitive set of size $\kappa$ satisfying $\mathsf{ZFC}^-$ with $\kappa \in M$.
(2) For weak $\kappa$-model $M$, an $M$-ultrafilter $U$ on $\kappa$, is 0-good if the ultrapower of $M$ by $U$ is well-founded.
(3) For weak $\kappa$-model $M$, an $M$-ultrafilter $U$ on $\kappa$ is weakly amenable if for every $A \in M$ of size $\kappa$ in $M$, the intersection $U \cap A$ is an element of $M$.
(4) For weak $\kappa$-model $M$, an $M$-ultrafilter on $\kappa$ is 1-good if it is 0-good and weakly amenable.

**Fact C.13** ([18]) *Suppose $M$ is a weak $\kappa$-model, $U$ is a $\alpha$-good $M$-ultrafilter on $\kappa$, and $j : M \to N$ is the ultrapower by $U$. Define $j(U) = \{A \in \mathscr{P}(j(\kappa))^N \mid A = [f]$ and $\{\alpha < \kappa \mid f(\alpha) \in U\} \in U\}$. Then $j(U)$ is a weakly amenable $N$-ultrafilter on $j(\kappa)$ such that $j``U \subseteq j(\kappa)$.*

Suppose $M$ is a weak $\kappa$-model and $U_0$ is a 1-good $M$-ultrafilter on $\kappa$. From Fact C.13, let $j(U_0) = U_1$ be the weakly amenable ultrafilter for the ultrapower of $M$ by $U$. If the ultrapower by $U_1$ happens to be well-founded, we will say that $U_0$ is 2-good.

In this way, we can continue iterating the ultrapower construction so long as the ultrapowers are well-founded. For $\alpha \leq \omega$, we will say that $U$ is $\alpha$-good if the first $\alpha$-many ultrapowers are well-founded. Suppose next that the first $\omega$-many ultrapowers are well-founded. We can form their direct limit and ask if that is well-founded too.

If the direct limit of the first $\omega$-many iterates turns out to be well-founded, we will say that $U$ is $\omega + 1$-good. Continuing the pattern, we make the following definition.

**Definition C.14** ([18]) Suppose $M$ is a weak $\kappa$-model and $\alpha$ is an ordinal. An $M$-ultrafilter on $\kappa$ is $\alpha$-good, if we can iterate the ultrapower construction for $\alpha$-many steps.

**Fact C.15** ([18]) *Suppose $M$ is a weak $\kappa$-model. An $\omega_1$-good $M$-ultrafilter is $\alpha$-good for every ordinal $\alpha$.*

From Fact C.15, to iterate the ultrapower construction through all the ordinals, it suffices to know that we can iterate through all the countable ordinals. Thus, the study of $\alpha$-good ultranlters only makes sense for $\alpha \leq \omega_1$.

**Definition C.16** ([18]) For $\alpha \leq \omega_1$, a cardinal $\kappa$ is $\alpha$-iterable if every $A \subseteq \kappa$ is contained in a weak $\kappa$-model $M$ for which there exists an $\alpha$-good $M$-ultrafilter on $\kappa$.

**Proposition C.2** ([18])

(1) If $\kappa$ is an $\alpha$-iterable cardinal$\alpha$-iterable cardinal, then the cardinal $\kappa$ is a limit of $\beta$-iterable cardinal for $\beta < \alpha$;
(2) For $\alpha < \omega_1^L$, the $\alpha$-iterable cardinals$\alpha$-iterable cardinal are downward absolute to $L$. This result is optimal since $\omega_1$-iterable cardinals cannot exist in $L$;
(3) If $\kappa$ is a remarkable cardinal, then there is a countable transitive model of ZFC with a proper class of 1-iterable cardinals.
(4) If $\kappa$ is 2-iterable, then $\kappa$ is a limit of remarkable cardinal;
(5) If there is an $\omega_1$-iterable cardinal, then $0^\sharp$ exists;
(6) If $0^\sharp$ exists, then the Silver indiscernibles are $\alpha$-iterable in $L$ for all $\alpha < \omega_1^L$;

The notion of $\alpha$-Erdös cardinal is introduced next. For definitions and facts about $\alpha$-Erdös cardinal, I refer to [5, 6].

**Definition C.17** ([5, 6]) For any limit ordinal $\alpha$, $\alpha$-Erdös cardinal is the least cardinal $\kappa$ such that $\kappa \to (\alpha)_2^{<\omega}$.

**Fact C.18** ([5, 18])

(1) For $\omega \leq \alpha < \omega_1^L$, $\alpha$-Erdös cardinal are downward absolute to $L$.
(2) An $\omega$-Erdos cardinal implies for every $n \in \omega$, the consistency of the existence of a proper class of $n$-iterable cardinals.
(3) The existence of $\omega_1$-Erdös cardinal implies $V \neq L$.

Let us consider some equivalences involving "$0^\sharp$ exists" and some consequences of "$0^\sharp$ does not exist" in ZFC.

**Theorem C.19** ([5, 6, 19]) *The following statements are equivalent:*

*(1)* $0^\sharp$ *exists.*
*(2)* *There exists an elementary embedding from L to L.*
*(3)* *For some $\alpha$ and $\beta$, there exists $j : L_\alpha \prec L_\beta$ with $crit(j) < |\alpha|$.*
*(4)* *There exists an L-ultrafilter U such that the ultrapower of L by U is well-founded.*
*(5)* *There exists an iterable L-ultrafilter.*
*(6)* *For some limit ordinal $\lambda$, $(L_\lambda, \in)$ has uncountable set of indiscernibles.*
*(7)* *There exists a class I of ordinals such that I is a closed unbounded class of indiscernibles for L and the Skolem hull of I in L is L (elements of I are called silver indiscernibles).*[6]
*(8)* *(A)* *If $\kappa, \lambda$ are uncountable cardinal and $\kappa < \lambda$, then $(L_\kappa, \in) \prec (L_\lambda, \in)$.*
    *(B)* *There exists a unique class sized club of ordinals I containing all uncountable cardinals such that for any uncountable cardinal $\kappa$:*
        *(a) $|I \cap \kappa| = \kappa$;*
        *(b) $I \cap \kappa$ is a set of indiscernibles for $(L_\kappa, \in)$;*
        *(c) Any $a \in L_\kappa$ is definable from $I \cap \kappa$ in $(L_\kappa, \in)$.*
*(9)* $\aleph_\omega$ *is regular in L.*
*(10)* *Any $X \in \mathscr{P}(\kappa) \cap L$ either contains or is disjoint from a closed and unbounded subset of $\kappa$ where $\kappa$ is an uncountable regular cardinal.*
*(11)* *For any $\Sigma^1_1$ game, either player I has a winning strategy or player II has a winning strategy which is recursive in $0^\sharp$.*

**Theorem C.20** ([5, 6, 19]) *If $0^\sharp$ does not exist, then:*

*(1)* *Any singular cardinal is a singular cardinal in L.*
*(2)* *For any singular cardinal $\kappa$, $(\kappa^+)^L = \kappa^+$.*
*(3)* SCH *holds. i.e. for singular cardinal $\kappa$, if $2^{cf\kappa} < \kappa$, then $\kappa^{cf\kappa} = \kappa^+$.*
*(4)* *Let $\kappa$ be a singular cardinal, if $\forall \alpha < \kappa (\mathscr{P}(\alpha) \subseteq L)$, then $\mathscr{P}(\kappa) \subseteq L$.*
*(5)* *(Covering theorem) For any uncountable set X of ordinals, there is a $Y \in L$ such that $X \subseteq Y$ and $|Y| = |X|$.*

In this book, $\kappa$-model is a model in the form $L[U]$ such that $\langle L[U], \in, U \rangle \models U$ is a normal ultrafilter over $\kappa$. For the theory of $0^\dagger$, I refer to [6].

**Theorem C.21** (Solovay, [6]) *The following are equivalent:*

*(1)* $0^\dagger$ *exists.*
*(2)* *There exists a $\kappa$-model $\mathscr{M}$ for some ordinal $\kappa$ that has an uncountable set I of indiscernibles such that $\min(I) > \kappa$.*

---

[6]Moreover, with $\langle \tau_\xi \mid \xi \in \mathrm{Ord} \rangle$ the increasing enumeration of $I$, then: (1) $I$ contains every uncountable cardinal; (2) $|\tau_\xi| = |\xi| + \aleph_0$; (3) for any limit ordinal $\alpha \geq \omega$, the Skolem hull of $\{\tau_\xi \mid \xi < \alpha\}$ in $L_{\tau_\alpha}$ is $L_{\tau_\alpha}$; (4) if $\xi \leq \eta$, then $L_{\tau_\xi} \prec L_{\tau_\eta} \prec L$.

(3)  *For every uncountable cardinal $\lambda$, there exists a $\lambda$-model $\mathcal{M}$ and a double class $\langle X, Y \rangle$ of indiscernibles for $\mathcal{M}$ such that:*

    (a)  *$X \subseteq \lambda$ is closed and unbounded;*
    (b)  *$Y \subseteq \mathrm{Ord} \setminus (\lambda + 1)$ is a closed unbounded class;*
    (c)  *$X \cup \{\lambda\} \cup Y$ contains every uncountable cardinal;*
    (d)  *The Skolem hull of $X \cup Y$ in $\mathcal{M}$ is $\mathcal{M}$.*

**Definition C.22**  A cardinal $\kappa$ is *Woodin* if for all $A \subseteq \kappa$ there are arbitrarily large $\delta < \kappa$ such that for all $\alpha < \kappa$ there is some elementary embedding $\pi : V \to M$ with $M$ transitive and with critical point $\delta$ such that $V_\alpha \subseteq M$ and $A \cap \alpha = \pi(A) \cap \alpha$.

**Definition C.23**  Define $M_n^\sharp(x) = \langle L_\gamma[E, x], \in, x, E, U \rangle$ where

- $U$ is an $L[E, x]$-ultrafilter;
- $E$ is a sequence of extenders (a system of ultrafilters) witnessing that $L_\gamma[E, x] \models$ there are $n$ Woodin cardinals;
- $M_n^\sharp(x)$ is iterable.[7]

"$M_n^\sharp(x)$ exists" is a large cardinal axiom which, for $n > 0$, is much stronger than "$0^\sharp$ exists". "$M_0^\sharp(\emptyset)$" is just $0^\sharp$.

# References

1.  Sami, R.L.: Analytic determinacy and $0^\sharp$: A forcing-free proof of Harrington's theorem. Fundamenta Mathematicae **160** (1999)
2.  Mansfield, R., Weitkamp, G.: Recursive Aspects of Descriptive Set Theory. Oxford University Press, Oxford (1985)
3.  Harrison, J.: Recursive pseudo-well-orderings. Trans. Amer. Math. Soc. **131**, 526–543 (1968)
4.  Moschovakis, Y.N.: Descriptive Set Theory. North-Holland, Amsterdam (1980)
5.  Jech, T.J.: Set Theory. Third Millennium Edition, revised and expanded. Springer, Berlin (2003)
6.  Kanamori, A.: Higher Infinite: Large Cardinals in Set Theory from Their Beginnings, 2nd Edn. Springer Monographs in Mathematics, Springer, Berlin (2003)
7.  Sacks, G.E.: Higher Recursion Theory. Perspectives in Mathematical Logic, Springer (1990)
8.  Harrington, L.A.: Analytic determinacy and $0^\sharp$. J. Symb. Log. **43**, 685–693 (1978)
9.  Weitkamp, G.: Analytic sets having incomparable Kleene degrees. J. Symb. Log. **47**(4), 860–868 (1982)
10.  Dubose, D.A.: The Equivalence of Determinacy and Iterated Sharps. J. Symb. Log. **55**(2), 502–525 (1990)
11.  Kunen, K.: Set Theory: An Introduction to Independence Proofs. North Holland (1980)
12.  Mekler, A.H., Saharon, S.: The consistency strength of "every stationary set reflects". Israel J. Math. **67**(3), 353–366 (1989)
13.  Johnstone, T.A.: Strongly unfoldable cardinals made indestructible. J. Symb. Log. **73**(4), 1215–1248 (2008)
14.  Hamkins, J.D., Džamonja, M.: Diamond (on the regulars) can fail at any strongly unfoldable cardinal. Ann. Pure Appl. Logic. **144**(1–3), 83–95 (2006)

---

[7]For the definition of iterability, I refer to Sect. 2.1.2.

15. Villaveces, A.: Chains of end elementary extensions of models of set theory. J. Symb. Log. **63**(3), 1116–1136 (1998)
16. Baumgartner, J.: Ineffability properties of cardinals I. In: Hajnal, A. et al. (Eds.), Colloq. Math. Sot. Janos Bolyai 10, Infinite and Finite Sets, Vol. III, pp. 109–130. North-Holland, Amsterdam (1973)
17. Baumgartner, J.: Ineffability properties of cardinals II. In: Butts and Hintikka (Eds.), Logic, Foundations of Mathematics and Computation Theory, pp. 87–106. Reidel, Dordrecht (1977)
18. Gitman, V., Welch, P.: Ramsey-like cardinals II. J. Symb. Log. **76**(2), 541–560 (2011)
19. Devlin, K.J.: Constructibility. Springer, Berlin (1984)

# Index

© The Author(s), under exclusive license to Springer Nature Singapore Pte Ltd. 2019
Y. Cheng, *Incompleteness for Higher-Order Arithmetic*, SpringerBriefs in Mathematics, https://doi.org/10.1007/978-981-13-9949-7

Printed in the United States
By Bookmasters